Structure and Functions of White Clover Root Systems

Structure and Functions of White Clover Root Systems

Editor
Dr. P.B. Bansal

2007
Gene-Tech Books
New Delhi - 110 002

ISBN 81-89729-21-7

Published by : **GENE-TECH BOOKS**
4762-63/23, Ansari Road, Darya Ganj,
NEW DELHI - 110 002
Phone: 41562849
e-mail: genetechbooks@yahoo.co.in

Printed at : **Tarun Offset Printers**
Delhi

PRINTED IN INDIA

Preface

White clover is a species of clover native to Europe, North Africa, and West Asia. It has been widely introduced elsewhere in the world as a pasture crop. It is smooth-stemmed, with trifoliate leaves. The leaflets are oval or circular; their edges have tiny serrations, and there are whitish markings on the middle of the upper surface. The seedlings initially grow a tap root and then stolons arising from leaf axils form a branched, radiating network, colonizing bare soil spaces. When the short-lived seedling plant dies, the stems arising from rooted stolons grow independently. White clover improves soil quality, nourishes beneficial insects, and can be used in pastured animal systems.

White clover grows in turfgrass, crops, and landscapes. It is also found in a wide range of different field type environments. It can grow on many different types and pHs of soil, but prefers clay. White clover is relatively shallow rooted, with most roots in the top 20 cm of soil, although some roots extend to a depth of 1 m or more; clones expand through the adventitious rooting of stolons. The plant's extensive root system makes the soil more friable, improving tilth and water infiltration.

This book elaborately describes the structure and functions of white clover root systems. It will be highly beneficial to students, teachers and researchers of plant science and allied fields.

Editor

Contents

1

White Clover

White Clover (*Trifolium repens*) is a species of clover native to Europe, North Africa, and West Asia. It has been widely introduced elsewhere in the world as a pasture crop. It is a herbaceous perennial plant. It is a low growing, with heads of whitish flowers, often with a tinge of pink or cream. The heads are generally 1.5-2 cm wide, and are at the end of 7 cm peduncles or flower stalks. The leaves are trifoliolate, smooth, elliptic to egg-shaped and long-petioled. The stems function as stolons, so white clover often forms mats with the stems creeping as much as 18 cm a year, and rooting at the nodes.

Flower of White Clover

Cultivation and Uses

White clover grows in turfgrass, crops, and landscapes. It is also found in a wide range of different field type environments. White clover can tolerate close mowing. It can grow on many different types and pHs of soil, but prefers clay. Besides making an excellent forage crop for livestock, clovers are a valuable survival food: they are high in protein, widespread, and abundant. They are not easy to digest raw, but this can be easily fixed by boiling for 5-10 minutes. Dried flowerheads and seedpods can also be ground up into a nutritious flour and mixed with other foods. Dried flowerheads also can be steeped in hot water for a healthy, tasty tea-like infusion.

Before the introduction of broad-leaf herbicides, white clover was often added to lawn seed mixes, as it is able to grow and provide green cover in poorer soils where turfgrasses do not perform well. It is now generally considered a weed when growing in lawns.

Cultivar

White clover cultivars are classified arbitrarily by size of the plants, as small, intermediate, and large. Small varieties frequently include the words "Wild White" in their names. Most common are the intermediate-sized white clovers, which behave as annuals when stressed. Large type white clovers do well where climatic and soil conditions are optimal but do not have as wide environmental tolerances as the smaller types. Of the three types, small is considered a weedy type.

Ladino clover is a variety of white clover; intermediate types of white clover are more tolerant of heat than are ladino-type white clovers. There are Canadian varieties of white clover that tolerate soil pH as low as 4.5.

Seeds

Seeds ovoid-truncate, ca 1.5 mm long and wide, surface smooth, usually entirely yellowish, sometimes reddish; 1,764,000 seeds per kg. White clover seed are small (1,764,000/kg, 77 kg/hl). White clover shows strong seedling vigor. Seedlings are epigeal, and germination occurs at the soil surface.

Mature Plant

White clover usually behaves as a perennial, but it performs as an annual in warm climates and a biennial in cold. Stolons glabrous, solid, creeping, 1-4 dm long, leaves long-petioled; leaflets cuneate-obovate to broadly oblong, emarginate to obtuse, 1-3 cm long, 0.75-2.5 cm broad, solid dull green or occasionally marked with a white "V" and sometimes with dark red flecks, denticulate; stipules pale except for the conspicuous nerves, abruptly narrowed to a point; peduncles scapiform, usually very elongate; heads subglobose in fruit, rather loose, 1.5-2.0 cm in diameter; flowers 20-40 (100); bracts membranous, at first equaling the pedicels but much exceeded by them as the latter elongate; pedicels often sparsely puberulent, elongating and recurving in age, the inner longer than the outer; flowers 6-10 mm long, fragrant; calyx-teeth unequal, the upper slightly shorter than the tube, the lower two-thirds its length; corolla white or rose tinged, becoming brown, 8 mm long, twice the length of the calyx; pods sessile, linear, exserted, usually 3- to 4-seeded, with 75-100 seeds per head.

As a perennial legume, this species spreads by stolons (Slayback, pers. comm.). Leaves, stems, and flower heads of Ladino clover generally are from two to four times as large as common white clover. In shape, colour and markings, the leaves and flower heads look like common white clover. White clover can develop a taproot 1 m deep, but it dies at the end of the first year, and secondary roots developing from the stolon become the main root system. The roots of white clover are mainly shallow.

Temperature

White clover tolerates mean annual temperatures of 4.3-21.8 C, with the mean of 101 cases being 11.6. White clover grows best under cool, moist conditions and is about as hardy as Red Clover or Alfalfa. Hofstetter termed it extremely winter-hardy. Miller *et al.* advised that in districts subject to severe frosts, clover-seed fields should enter winter with at least 6 to 8 inches of growth. Such growth protects the stand against excessive frost damage. With mild climates, the grazing can continue into winter without serious damage ensuing.

Heritability of frost tolerance in white clover is high (0.75-0.93), suggesting that incorporation of frost tolerance into agronomically-suitable, but frost-sensitive cultivars could be successful. As heat tolerance is concerned, intermediate types of white clover are more resistant than ladino types. Hot dry periods during the summer reduce growth of white clover, but plants will usually survive.

Geographic Range

White clover probably originated in the eastern Mediterranean region of Asia Minor, and was spread to other continents by colonists; it ranges from the boreal moist to wet through subtropical dry to rain forest life zones. It is naturalised to wet places in California, especially in the mountains and in northern parts of the state. The species grows in a wide range of habitats, including dry meadows, mudflats, rarely on saline meadows, wood margins, open woods, banks of rivers and brooks, plains, semi-desert regions, and mountains up to the subalpine meadows. It is a frequent weed on roadsides and in barrens.

Miller *et al.* noted that seed production is largely confined to irrigated areas in California, Oregon, and Idaho, where weather is favourable. Seed can probably be produced in most of California at elevations under 3,000 feet, except where summer heat is excessive. High winds at harvest (July, August, and September) can be a problem. Excessively fog or rain and snow at harvest may disrupt harvest and reduce seed quality.

Water

White Clover grows best under cool, moist conditions and is less drought tolerant than trefoil. Miller *et al.* wrote that seed production usually necessitates irrigation about every 7 to 10 days during summer. With frequent irrigations, to produce good seed on open soils may require from 2 to 3 times that needed on heavier type soils. On heavy clay soils the irrigation interval may range from ten to twenty-one days. Miller *et al.* recommended the strip-check system of land preparation and irrigation.

A modified version of the contour check system of irrigation, was also successful for irrigated pastures but was rarely used for Ladino seed production. Immediate and complete drainage following the irrigation is essential to avoid seed damage through germination and drowning of plants. Sprinklers may be used where land cannot be levelled. Clover-seed fields require frequent irrigation, so land preparation should be calculated to minimise labour costs.

Madson suggested that white clover should be irrigated when leaves begin to cup together. This may occur in sandy or hard spots at first. Field condition in general should be considered. Do not withhold irrigation until the entire field is wilting. Irrigation should replenish the soil moisture to the depth of the white clover roots. Examine the soil to ascertain the depth of soil and the root zone. From 1 to 3 inches of water per irrigation will usually suffice. Larger applications will be wasteful and perhaps harmful. Ponding in hot weather can drown plants.

It is customary to withhold water for 4 to 10 days before a field is cut for seed. Miller *et al.*, however, advised not intentionally drying up a field which has a good developing seed crop, because this would reduce seed yield, cause much seed to shrivel, and reduce germination. Whereas Ladino clover requires frequent irrigation, its annual requirement usually does not exceed that of alfalfa. In most of California, Ladino requires from 3 to 5 acre-feet of water per acre annually. In the interior valleys on open, permeable soils 6 to 9 acre-feet year may be needed.

If water is withheld before harvest, irrigate immediately afterward, to minimise drought damage. This can also increase fall pasturage. Irrigations should be given frequently in the fall until the seasonal rains begin. White clover can grow on clays to silty loams if water is available all year. Ladino clover does well on soils with a high water table or with poor drainage, but it is not suited to dry, swampy, high-alkaline, or highly-saline sites. Mixtures of grasses and white clover are difficult to maintain because grasses may shade the clover and because the clover requires high soil moisture levels. If soil nitrogen is high, grasses will tend to predominate, whereas the clover will dominate if nitrogen is low.

Nutrients

White clover may require the equivalent of 300-600 kg/ha of 20% superphosphate and 60-250 kg/ha of muriate of potash. It is a poor competitor with grass for soil P. 112 white clover cultivars and 7 breeding lines from 25 countries were evaluated for response to added P. Cultivars that responded to low levels of added P were all from countries with cold winters. 119 cultivars and lines of white clover were originally evaluated for response to P addition. Based on principal components analysis, these were assigned to 8 groups. 26 of the varieties were selected to represent the 8 groups. The varieties that absorbed the most P also accumulated the most N.

Caradus *et al.* wrote that in New Zealand, the large-leaved white clover cultivar 'Grasslands Kopu' showed better growth response to P and higher yields than did 'Barbian', 'Grasslands Tahora', or 'Grasslands Huia.' Maximum yield of 'Grasslands Kopu' was also attained at a lower level of added P. White clover responds well to addition of gypsum or flowers of sulfur as sulfur sources. It showed a 63% increase in dry matter production following application to a 2-year-old stand.

Mixtures of grasses and white clover are difficult to maintain because grasses may shade the clover and because the clover requires high soil moisture levels. If soil nitrogen is high, grasses will tend to predominate, whereas the clover will dominate if nitrogen is low. Kanyama-Phiri *et al.*, wrote that in strawberry clover—white clover—perennial ryegrass—orchardgrass mixtures, the clovers were most productive with 60 kg/ha of N added. Strawberry clover and ladino clover are similar in morphology and growth habit but differ in response to different nutrient and grazing regimes. In strawberry clover—white clover—perennial ryegrass—orchardgrass mixtures, with no added nitrogen, ladino clover peaked in dry matter production in October, then fell off during November. Strawberry clover peaked in November. By contrast, when nitrogen was added (60, 120, and 180 kg/ha.of N added), these situations were reversed.

Soil pH

Duke wrote that white clover tolerates soil pH of 4.5-8.2, with the mean of 89 cases being 6.3, and stated that the species or cultivars thereof show tolerance to both high and low pH. The species is said not to grow well in alkaline soil, and the best pH range is 6.0 to 6.5, 6 to 7, or 6.5 to 7.0, with the optimal pH for white clover said to be 6.5. Some Canadian varieties of white clover can tolerate soil pH as low as 4.5, but acid soil should be limed to a pH of 6.0-6.5 before seeding white clover. White clover is not suited to dry, swampy, high-alkaline, or highly-saline sites.

Duke wrote that white clover or varieties thereof tolerate aluminum, drought, heavy soil, high ph, low pH, poor soil, and salt. The species does better on clay or loam soils than on sandy soils. White clover grows best on well-drained loam or clay soils of high fertility. McLeod advised that heavy, moist, well-drained, highly fertile soils are best and stated that with abundant rainfall, white clover will grow on sandier soil, but not reliably. Ladino clover does well on soils with a high water table or with poor drainage, but it is not suited to dry, swampy, high-alkaline, or highly-saline sites.

White clover can grow on clays to silty loams if water is available all year. When grown for seed production, heavier clay or loam soils are best. White clover does well on shallow soil underlain by a tight clay layer or a hardpan, even if this is 18 inches or slightly less of the surface. On deep, open, friable, fertile soils, seed production seldom is good because plants do not produce abundant seed heads. Luxuriant vegetative growth, sparse production of seed heads, and excessive water losses combine to reduce seed production.

Shade tolerance

In California, white clover is often included in mixtures with strawberry clover because the former is shade tolerant and the latter tolerates heat. Thus, in an orchard understory with dappled shade and sun, the two species complement one another (Bugg, pers. comm.). White clover is shade tolerant and does well in mature, irrigated, swampy pecan orchards in southern Georgia (Bugg, pers. comm.).

Salinity tolerance

Duke ascribed salt tolerance to white clover or cultivars thereof, but the species is not suited to highly-saline sites, according to Gibson and Cope. This confirms the finding by Miller *et al.*, who wrote that the species does not do well on very saline soils. Rogers *et al.* rated cultivars of white clover for salinity tolerance as follows: 'Haifa' and 'Irrigation' > 'Tamar' and 'Ladino'. With moderate salinity, highest yields occurred with cultivars that yield the most under non-saline conditions. Varieties that showed little decline with increased salinity yielded poorly in non-saline conditions.

Herbicide sensitivity

Duke reported that the cultivar 'Regal Ladino' was selected for tolerance to the herbicides 2,4-D, 2,4-DB, and 2,4,5-T. The herbicide 2,4-DB does not injure grass or white clover. Also, the herbicides propyzamide and carbetamide can be useful in suppressing incumbent grasses during the establishment phase of slot-seeded white clover.

Life Cycle

White clover usually behaves as a perennial but performs as an annual in warm climates and a biennial in cold. Stands regenerate themselves by reseeding and by vegetative (clonal) growth; both processes operate in the same stands. Perenniation is by continual re-establishment of seedlings or rooting of new stolons. Flowering and fruiting can occur nearly year-round in some areas, and spring to fall in the North. White clover flowers develop once days lengthen to 14 hours and 15 minutes. Flowers are insect pollinated, and abundant bees are essential to cross pollination and high seed production. White clover seedlings are epigeal; germination occurs at the soil surface.

Seeding Rate

Proposed seeding rates are variable: 0.45-5 kg/ha; 4 kg/ha; 5 lb/acre; 4 to 6 lb/acre; 4 lbs/acre by ground or aerial application; or 5 or 6 lbs/acre.

Seeding Depth

White clover seed should be sown at or near the soil surface, or at a depth of no more than 4 mm or 5 mm.

Seeding Method

Drilling is the preferred method, but broadcasting will suffice. Miller *et al.* advised that, in preparing to seed white clover for seed production, it is desirable to have a fine, firm seedbed with a firm, moist bottom covered by two to three inches of moist, soil free of fine tilth.

Cultipack the soil before seeding. Loose, cloddy seedbeds filled with air pockets may lead to patchy, poor stands. Before planting, an irrigation settles and fills, revealing irregularities that can be corrected and provides moisture for germinating seed. After the irrigation, level the field if irregular settling has occurred, then harrow and seed.

In establishing clover simultaneously with grass, drill grass seed in rows 40 to 50 cm apart, then broadcast white clover seed and cover them using a cultipacker. If establishing white clover amid preexisting warm-season grass sod, the clover should be seeded (possibly along with a cool-season grass such as ryegrass) in the fall when the grass is dormant. Mow or graze the grass closely, broadcast the inoculated clover seed, and use a roller to press it into the soil. If establishing the white clover during the spring, grass competition should be reduced by mowing or grazing closely, or by damaging strips of the grass by ploughing or banding herbicide. The herbicides propyzamide and carbetamide can be useful in suppressing incumbent grasses during the establishment phase of slot-seeded white clover.

Seeding Dates

Depending on local conditions, seeding should be done between September 15 and November 15. Planting in the early fall insures larger and more productive Ladino plants in the first crop year. Spring seedings are preferable if fields are foul with winter weeds, although the chances of obtaining a satisfactory seed crop are fewer. In areas subject to severe winter frosts, planting should be completed early enough in the fall to insure seedling establishment before the first hard freeze.

Inoculation

White clover requires rhizobial inoculant type "B". In

reestablishing stands in warm regions, white clover seed should always be inoculated before sowing. In cool regions, rhizobia from the preexisting white clover crop will survive in the soil.

Seed Cost

Dutch white clover cost $2.80 per lb, New Zealand white clover $2.80, and ladino clover $3.33 retail in 1993.

Days to Flowering

The flowering period in California as April through December. According to Duke, flowering can be nearly year round in some areas and is from spring to fall in the North. White clover shows a series of flowering periods or seed sets, which mature successively. Blooming periods may be 12 to 20 days apart and often overlap. Maximum bloom usually occurs 4 to 6 weeks after the last spring grazing or cutting.

Days to Maturity

Seed ripens in July and August in the Central Valley.

Seed Production

White clover seed is produced mainly in irrigated areas of the West. Duke summarised seed yields as being 30-200 kg/ha in humid climates and 150-600 kg/ha in the West. First-season seed yields of 100-350 lbs/acre may result from early fall planting. Bees are important in pollination and seed production, and should be protected. Animal manures, unless known to be weed-seed free, should not be used on seed fields.

For seed formation, Ladino clover flowers must be cross-pollinated. 1 to 1.5 beehives per acre is sufficient for complete pollination. Bees should be moved in when blossoming starts, usually 2 to 3 weeks after the spring grazing or cutting is finished. In California, bees are usually moved in about May 10, and are retained through the blooming period, i.e., until late August or September. The average production of Ladino clover seed in California is about 100 lbs/acre. Seed yields vary tremendously depending upon management and age of stand. Early fall-planted, well-managed stands may produce 100 to 300 lbs or more of seed per acre the first season.

Some spring-planted stands do not produce any seed at all the first year. Second-year stands give peak seed production, and may yield 250 to 5000 lbs per acre. After the third year, seed yields usually decline rapidly, and there is seldom production after the fifth year. Weeds such as ryegrass cause declining seed yields from the third year on.

Growth Habits of White Clover

White clover usually behaves as a perennial, but it performs as an annual in warm climates and a biennial in cold. The primary axis seldom lasts more than 2 years, but plants persist by forming new growth centers through rooted stolons and their apical meristems. The plants are shallow rooted and spread by stolons that root at the nodes. The species acts as a winter annual under adverse conditions, as in the southern United States.

The species is low growing with prostrate stems. White clover grows rapidly, and is a long-lived, true clover, spreading by creeping stems called stolons. These elongate rapidly and root at the joints if soil is moist. Thus, sparse stands may thicken by the end of the first year. As related by Turkington, *Trifolium repens* can reproduce either vegetatively or by seed. In a permanent pasture, the species may display phenotypic plasticity allowing a clone to respond to local conditions (fine-grained environmental variability), and genotypic variability allowing different genets to respond differentially to coarse-grained environmental variability.

Maximum Height

Height has been variously reported as 8-10 inches on good sites or 5-30 cm. Mixed stands of strawberry and white clovers attained height of 33.0 +/- 3.3 cm (mean +/- SEM) in early May, following October seeding in Mendocino County, California.

Root System

White clover is relatively shallow rooted, with most roots in the top 20 cm of soil, although some roots extend to a depth of 1 m or more; clones expand through the adventitious rooting of stolons. Miller confirmed that the roots of white clover are mainly shallow; it can develop a taproot 1 m deep, but it dies at the end of the first year, and secondary roots developing from the stolon become the main root system.

Miller *et al.* wrote that white clover plants have most roots in the upper 18 to 24 inches of soil. Therefore, Ladino clover does well on soils too shallow for deep-rooted crops like alfalfa, sugar beets, or tree crops. On deep, open soils, roots extend to four or five feet. White clover taproot can extend to a depth of 1-3 ft. Kutschera reported that white clover generally roots to a depth of 61-76 cm.

Establishment

Ladino is slow to establish. Keep surface soil moist during germination. Early fall and late spring seedings require frequent light irrigations timed to keep the surface soil moist, preventing crusting and enabling emergence. Fall seeding is favoured because rains may help establishment. Do not allow the soil surface to remain dry more than two days until clover plants have at least four true leaves.

White clover clones expand through the adventitious rooting of stolons; nonetheless, establishment is crucial. A good initial stand of white clover outperforms poor stands even after 3 years. The herbicides propyzamide and carbetamide can be useful in suppressing incumbent grasses during the establishment phase of slot-seeded white clover. White clover behaves as a perennial by continual re-establishment of seedlings or rooting of new stolons. White clover stands regenerate themselves by reseeding and by vegetative (clonal) growth; both processes operate in the same stands.

White clover clones expand through the adventitious rooting of stolons; nonetheless, establishment is crucial, and a good initial stand of white clover outperforms poor stands even after 3 years. Young or thin stands can thicken rapidly. When foliage is removed by grazing or clipping, recovery occurs in from 21 to 28 days.

White clover performs best under regimes of heavy grazing or mowing. Heavy accumulation of vegetation is deleterious to white clover because light is reduced, soil moisture may be depleted, and the microclimate predisposes for disease and insect problems. Under such a regime, white clover is prone to have very succulent growth, and is unlikely to survive the microclimatic change that would result from sudden heavy mowing or grazing.

In managing mixed stands, grass height must be controlled to allow white clover to do well. In mixtures of white clover and bermuda grass or another warm- season grass, nitrogen should be applied in the late spring or summer, if the desire is to encourage the grass. If the desire is to encourage cool-season grasses, the nitrogen should be applied in the fall or early spring.

Mixtures of grasses and white clover are difficult to maintain because grasses may shade the clover and because the clover requires high soil moisture levels. If soil nitrogen is high, grasses will tend to predominate, whereas the clover will dominate if nitrogen is low. In Austria, Danso *et al.* conducted a 2-year trial in a triple-species mixed sward of white clover (cv *'Zapican'*), birdsfoot trefoil (*Lotus corniculatus* cv *'Gabriel'*) and fescue (*Festuca arundinacea* cv *'Tacuabe'*).

White clover showed good production for the first harvest of the first year; thereafter, birdsfoot trefoil dominated. In the first year, both legumes contribute about equally to the approximately 130 kg N/ha fixed in the sward. In the second year, white clover only contributed 5% of the 46 kg N/ha fixed in the last two harvests. Mixtures including the two legumes have an advantage because the early production by white clover is complemented by later production and better persistence by birdsfoot trefoil. Stands with multiple legumes often show better livestock weight gains.

Kanyama-Phiri *et al.* reported that in strawberry clover—white clover—perennial ryegrass—orchardgrass mixtures, close mowing alternating with 30-31 day regrowth periods apparently favoured the two clovers and the perennial ryegrass. Kanyama-Phiri *et al.* asserted that Strawberry clover and ladino clover are similar in morphology and growth habit but differ in response to different nutrient and grazing regimes. In strawberry clover—white clover—perennial ryegrass—orchardgrass mixtures, with no added nitrogen, ladino clover peaked in dry matter production in October, then fell off during

November. Strawberry clover peaked in November. By contrast, when nitrogen was added, these situations were reversed.

Trifolium repens can reproduce either vegetatively or by seed. In a permanent pasture, the species may display phenotypic plasticity allowing a clone to respond to local conditions (fine-grained environmental variability), and genotypic variability allowing different genes to respond differentially to coarse-grained environmental variability.

Grazing or Mowing

White clover performs best under regimes of heavy grazing or mowing. In tolerating close grazing (and, presumably, mowing) ladino white clover is exceeded only by birdsfoot trefoil and subterranean clover. Heavy accumulation of vegetation is deleterious to white clover because light is reduced, soil moisture may be depleted, and the microclimate predisposes for disease and insect problems. Under such a regime, white clover is prone to have very succulent growth, and is unlikely to survive the microclimatic change that would result from sudden heavy mowing or grazing.

In managing mixed stands of white clover and grasses, grass height must be controlled to allow white clover to do well. Frequent mowing will encourage the clover, which revives quickly when cut for hay, because stems are not cut. Four or five cuttings per year, at 35-40 day intervals, can be harvested in some areas. Miller *et al.* advised that first-year stands should not be grazed, except near the end of the season; otherwise, severe injury to the stand can result. By contrast, spring grazing or mowing until mid-May is typical, helps in weed control weeds and promotes more uniform flowering and seed production.

According to Miller, white clover is subject to root damage by the fungi Sclerotium rolfsii and Sclerotinia trifoliorum. Root and stolon damage is caused by the fungi Rhizoctonia, Fusarium, Leptodiscus, Curvularia, and Colletotrichum. In the Southeast, the root-knot nematodes, Meloidogyne spp., damage white clover roots. Host plant resistance, crop rotation, soil fertility, and proper cutting schedules can aid in control of these pests.

As noted by Kanyama-Phiri *et al.*, in strawberry clover—white clover—perennial ryegrass—orchardgrass mixtures, close mowing alternating with 30-31 day regrowth periods apparently favours the two clovers and the perennial ryegrass. Strawberry clover and ladino clover are similar in morphology and growth habit but differ in response to different nutrient and grazing regimes. In strawberry clover—white clover—perennial ryegrass—orchardgrass mixtures, with no added nitrogen, ladino clover peaked in dry matter production in October, then fell off during November. Strawberry clover peaked in November. By contrast, when nitrogen was added.

Incorporation

White clover is a perennial plant, and is seldom incorporated as a part of normal cultural practice (Bugg, pers. comm.). A good stand should provide 100 lb of N per acre when plowed at bud stage.

Harvesting

Seed may be harvested by three methods: (1) direct combining, (2) combining from the windrow, or (3) stationary thresher.

Miller *et al.* recommended harvesting for seed when 90-95% of the visible heads are brown and the flower stems (seed stalks) have started to dry. In the Sacramento and San Joaquin Valleys, this would be in July and August. Withhold irrigation for 4 to 10 days prior to harvest. According to Duke, seed should be harvested when most seed heads are light brown, about 25-30 days after peak blossoming. Mow and cure in swaths or in small windrows and retrieve using combines with pickup attachments. Each handling causes some shattering, so minimise the number of operation. Complete drying of seeds may require artificial means or spreading and turning the covered seed. White clover revives quickly after mowing because stems are not cut. Four or five cuttings per year, at 35-40 day intervals, can be harvested in some areas.

Uses of White Clover

Duke regarded white clover as among the most nutritious and widely-used forage legumes, with a value difficult to estimate because of its tendency to volunteer; about half the irrigated or humid pasture contains some white clover. It is usually grown in mixes with grasses, it makes excellent feed, pasture, hay, and silage for livestock and poultry. Once established it serves excellently as a cover crop and in stabilizing soil and reducing erosion. The latter point was confirmed by Miller. White clover is widely used as an irrigated pasture plant in California and is best adapted to rotational pasture management. As a cover crop, it is suitable in orchards where a perennial ground cover can be maintained and in vineyards of the same description.

Mixtures

White clover is usually grown in mixtures with grasses; the latter can be either temperate or tropical grasses.

Examples include Bermudagrass, bromegrass, dallisgrass, Kentucky bluegrass, orchardgrass, tall fescue, or timothy. In mixtures with grasses, frequent mowing will encourage the white clover. Companion or nurse crops of barley or oat are not recommended for fall seedings, but may be used for spring-sown Ladino on tight soils which seal or cake. Nurse plants lead to better spring-sown stands by shading and reducing surface moisture loss. Therefore, white clover seedlings are better able to establish. Mow the nurse crop promptly once the clover has established, to reduce competition.

Gibson and Cope recommended that if establishing clover simultaneously with grass, one should drill grass seed in rows 40 to 50 cm apart, then broadcast white clover seed and cover them using a cultipacker. In establishing white clover amid preexisting warm-season grass sod, the clover should be seeded (possibly along with a cool-season grass such as ryegrass) in the fall when the grass is dormant. Mow or graze the grass closely, broadcast the inoculated clover seed, and use a roller to press it into the soil. If establishing the white clover during the spring, grass competition should be reduced by mowing or grazing closely, or by damaging strips of the grass by ploughing or banding herbicide.

The herbicides propyzamide and carbetamide can be useful in suppressing incumbent grasses during the establishment phase of slot-seeded white clover. By contrast, the herbicide 2,4-DB does not injure grass or white clover. In mixtures of white clover and bermuda grass or another warm-season grass, Gibson and Cope advised that nitrogen should be applied in the late spring or summer, if the desire is to encourage the grass. If the desire is to encourage cool-season grasses, the nitrogen should be applied in the fall or early spring.

Mixtures of grasses and white clover are difficult to maintain because grasses may shade the clover and because the clover requires high soil moisture levels. If soil nitrogen is high, grasses will tend to predominate, whereas the clover will dominate if nitrogen is low. White clover is a poor competitor with grass for soil P. Woledge wrote that mixtures of perennial ryegrass (Lolium perenne L.) and white clover (Trifolium repens L.) yield less than heavily fertilised monocultures of the former. Addition of small amounts of nitrogen in the spring leads to reduction of the proportions of clover in the stand.

There was no evidence that fertilised grass overtopped clover. Clover showed a smaller ratio of leaf area to total above-ground dry weight, but had a higher proportion of its lamina in the upper, well-lit strata. In Austria, Danso *et al.* conducted a 2-year trial in a triple-species mixed sward of white clover (cv *'Zapican'*), birdsfoot trefoil (Lotus corniculatus cv *'Gabriel'*) and fescue. White clover showed good production for the first harvest of the first year; thereafter, birdsfoot trefoil dominated.

In the first year, both legumes contributed about equally to the approximately 130 kg N/ha fixed in the sward. In the second year, white clover only contributed 5% of the 46 kg N/ha fixed in the last two harvests. Mixtures including the two legumes have an advantage because the early production by white clover is complemented by later production and better persistence by birdsfoot trefoil. Stands with multiple legumes often show better livestock weight gains.

Kanyama-Phiri et al. wrote that in strawberry clover—white clover—perennial ryegrass—orchardgrass mixtures, close mowing alternating with 30-31 day regrowth periods apparently favours the two clovers and the perennial ryegrass. The clovers were most productive with 60 kg/ha of N added. Strawberry clover and ladino clover are similar in morphology and growth habit but differ in response to different nutrient and grazing regimes. In strawberry clover—white clover—perennial ryegrass—orchardgrass mixtures, with no added nitrogen, ladino clover peaked in dry matter production in October, then fell off during November. Strawberry clover peaked in November. By contrast, when nitrogen was added (60, 120, and 180 kg/ha of N added), these situations were reversed.

Trifolium repens in the same 1-ha old field will diverge genetically through time depending on the grass species with which it is associated. Reciprocal transplant experiments have demonstrated differential success of *T. repens* from areas dominated by the grasses Lolium perenne or Agrostis capillaris to areas dominated by the original grass or other perennial grasses. The results suggested that, even within a field, white clover clones are genetically adapted to perform better amid stands of particular grasses.

In replicated pot and field trials, Hartl, W. grew wheat with undersown white clover, black medic, *Medicago lupulina* L., *Persian clover, Trifolium resupinatum* L., and various associated weeds. In both pot and field trials, biomass of weeds alive at the time of harvest was reduced substantially (50% or more reduction). The field trials showed reduction of wheat yield, as well, for black medic and Persian clover, but a non-significantly-increased yield when white clover, Trifolium repens, was used. White and Scott in New York found that yield of cereal rye was less affected by living mulch of white clover, 'Ladino' clover, or red clover than of crown vetch, birdsfoot trefoil or alfalfa.

Biomass

The mean proportion of dry weight of white clover comprised by N is 0.04; 225 kg N/ha added annually to pastures on these assumptions, an annual biomass production of 5.9 Mg/ha can be projected. In New Zealand, with a 10-month growing period, the N figure can supposedly be doubled, and thus, presumably, so can the biomass figure. By contrast, if we assume the mean crude protein content of 22% given by Duke and 6.25% of the protein being N, the projection is 16.36 Mg/ha, which seems inordinately high.

Dry matter yield of white clover should equal or exceed that of red clover (1,500 to 2,000 lb/acre). In mid-May, a first-year stand of a fall-sown mixed stand of strawberry and white clover yielded an above-ground dry biomass of 3,9+/-0.9 Mg/ha (Mean +/- S.E.M.) in an organic vineyard in Mendocino County, California. This figure is significantly less than that of the standing biomass for various winter-annual vetches, clovers, and medics sown in the same trial. With weeds including, the corresponding figure is 6.8+/-1.2 Mg/ha. White clover responds well to addition of gypsum or flowers of sulfur as sulfur sources. It showed a 63% increase in dry matter production following application to a 2-year-old stand.

N Contribution

White clover may yield as much as 225 kg/ha of nitrogen per year, or up to double this in New Zealand, where a 10-month growing season prevails. In Austria, Danso *et al*. conducted a 2-year trial in a triple-species mixed sward of white clover (cv *'Zapican'*), birdsfoot trefoil (*Lotus corniculatus* cv *'Gabriel'*) and fescue (*Festuca arundinacea* cv 'Tacuabe'). White clover showed good production for the first harvest of the first year; thereafter, birdsfoot trefoil dominated. In the first year, both legumes contributed about equally to the approximately 130 kg N/ha fixed in the sward. In the second year, white clover only contributed 5% of the 46 kg N/ha fixed in the last two harvests. Mixtures including the two legumes have an advantage because the early production by white clover is complemented by later production and better persistence by birdsfoot trefoil. Stands with multiple legumes often show better livestock weight gains.

First-year white clover mean nitrogen content is 81 lb/a, according to Brinton. Nitrogen content of white clover averages 3.8%. A good stand should provide 100 lb of N per acre when plowed at bud stage. Haley and Hogue assessed influence of type of ground cover on apple aphid, Aphis pomi DeGeer (Homoptera: Aphididae), and its predators in a young apple orchard. Four ground cover regimes were compared: (1) fall cereal rye (Secale cereale), herbicided in spring and summer; (2) a mixture of white clover (*Trifolium repens*) and grass; (3) herbicided tree-row strips and grassed-in alleys; and (4) woven black-plastic strips in the tree row and grassed alleys.

The trial was initiated at the beginning of the second year of the orchard. Few aphidophagous insects of interest (e.g., the predatory mirids *Deraeocoris brevis* and *Campylomma verbasci*, the predatory midge *Aphidoletes aphidimyza, lady beetles, hover flies*, or *lacewings*), were found in the ground covers. In the first year of the study, leaf nitrogen and aphid and predator densities were lower on trees with the white clover-grass mixture. These differences did not occur the second year. Terminal growth was particularly depressed for apple trees with understories of white clover and grass.

McNeill and Wood conducting studies in an environmental chamber with an atmosphere enriched by 15N indicated that white clover did not rapidly transfer nitrogen to associated ryegrass. The plants were mycorrhizal. Mackay *et al.* evaluated 119 cultivars and lines of white clover for response to P addition. Based on principal components analysis, these were assigned to 8 groups. 26 of the varieties were selected to represent the 8 groups. The varieties that absorbed the most P also accumulated the most N.

In a field study in Uruguay, Mallarino and Wedin found that N-fixation was more markedly depressed through the application of nitrogen fertiliser (100kg/ha) for birdsfoot trefoil than for white clover or red clover. In combination with tall fescue (Festuca arundinacea), year diculture stands with birdsfoot trefoil, white clover, and red clover led to respective annual total N fixation of 70-223, 42-200, and 49-277 kg N/ha. White clover-tall fescue mixes did not require legume domination in order to maximise total N fixation. Red clover gave the greatest N fixation among the legumes tested.

In Sweden, Marstorp and Kirchmann reported the following values for stems and leaves of several legumes harvested from the field 100 days after emergence. and later used as green manure:

	% Of Dry Matter			
Species	Carbon	Nitrogen	Lignin C	C/N
Red Clover	45.9	2.61	9.08	17.6
White Clover	39.5	3.10	10.26	12.7
Black Medic	42.9	3.12	10.88	13.8
Persian Clover	46.1	3.20	11.38	20.9
Berseem Clover	45.3	2.52	12.47	18.0
Subterranean Clover	42.6	3.02	9.69	14.1

Non-N Nutrient Contribution

In the first year, white clover mean phosphorus content is 5 lb/a, and potassium content, 35 lb/a.

Effects on Water

In Wales, Mytton *et al.* evaluated the differential effects on soil structure and water infiltration of monocultural swards of white clover and perennial ryegrass and a mixture of the two. The replicated study employed undisturbed soil cores (100 mm i.d. x 180 mm length; silt loam acid brown earth) extracted from the field and arranged in a greenhouse. White clover stands led to greater water infiltration, 43.1g/4h, vs. 32.9 for annual ryegrass, vs. 39.4 for mix. Apparent greater soil friability in the clover cores was not reflected in soil bulk density of % soil aggregates >2mm upon wet sieving. Total porosity was little changed, but air-filled (macroporosity) was as follows: clover > mix > perennial ryegrass.

Effects on Soil

White clover is very good for soil conservation because stolons provide ground cover that stabilises soil. In Wales, Mytton *et al.* evaluated the differential effects on soil structure and water infiltration of monocultural swards of white clover and perennial ryegrass and a mixture of the two. The replicated study employed undisturbed soil cores (100 mm i.d. x 180 mm length; silt loam acid brown earth) extracted from the field and arranged in a greenhouse. White clover stands led to greater water infiltration, 43.1g/4h, vs. 32.9 for annual ryegrass, vs. 39.4 for mix. Apparent greater soil friability in the clover cores was not reflected in soil bulk density of % soil aggregates >2mm upon wet sieving. Total porosity was little changed, but air-filled (macroporosity) was as follows: clover > mix > perennial ryegrass.

Effects on Livestock

In Austria, Danso *et al.* conducted a 2-year trial in a triple-species mixed sward of white clover (cv *'Zapican'*), birdsfoot trefoil (*Lotus corniculatus* cv 'Gabriel') and fescue (*Festuca arundinacea* cv 'Tacuabe'). White clover showed good production for the first harvest of the first year; thereafter, birdsfoot trefoil dominated. In the first year, both legumes contributed about equally to the approximately 130 kg N/ha fixed in the sward. In the second year, white clover only contributed 5% of the 46 kg N/ha fixed in the last two harvests. Mixtures including the two legumes have an advantage because the early production by white clover is complemented by later production and better persistence by birdsfoot trefoil. Stands with multiple legumes often show better livestock weight gains.

White clover is very palatable to livestock. In mixture with grasses and other legumes, it provides highly productive, nutritive pasture and is the basic legume in California's irrigated pasture. Miller *et al.* wrote that cut Ladino clover dries slowly; it is usually ensiled rather than cured for hay. 70 to 80 lbs of cane molasses can be added per ton of green clippings going into the silo; this improves silage quality and feeding value.

Pest Effects on Insects

Gibson and Cope noted that heavy accumulation of vegetation is deleterious to white clover because light is reduced, soil moisture may be depleted, and the microclimate predisposes for disease and insect problems. Under such a regime, white clover is prone to have very succulent growth and is unlikely to survive the microclimatic change that would result from sudden heavy mowing or grazing. Avoid insecticides other than sulfur if bees are working seed-producing stands of white clover.

Duke wrote that many arthropods attack white clover, including grasshoppers, clover leaf weevil (Hypera punctata), potato leafhopper (Empoasca fabae), meadow spittlebug (Philaenis spumarius), garden fleahopper (Halticus bractatus), spider mites, clover root curculio (Sitona hispidula), green june beetle (Cotinus nitida), lesser clover leaf weevil (Hypera nigrirostris), clover head weevil (Hypera meles), clover seed weevil (Microtrogus picirostris). white clover flower midge (Dasineura gentneri), lygus bugs, clover aphid (Nearctaphis bakeri), green cloverworm (Plathypena scabra), cutworms, lesser cornstalk borer (Elasmopalpus lignocellus), alfalfa weevil (Hypera postica), blister beetles (Epicauta spp.), and the fall armyworm (Spodoptera frugiperda).

Potato leafhopper (Empoasca fabae) is the most important insect pest of white clover. Other pests include meadow spittlebug, clover leaf weevil, alfalfa weevil, lygus bugs, clover seed weevil, and clover seed midge. Crop resistance, weed control, and crop rotation may be important in controlling some insect pests. Miller *et al.* wrote that four species of spider mite (called red spiders) can cause damage, beginning in early May. These include Atlantic mite, brown almond mite, Pacific mite, and the two-spotted mite. These mites are usually found on the undersides of the leaves, and infested leaves become cupped, yellowish and spotted. Heavily infested fields may appear dried-out and reddish-brown. Lygus bugs may build up rapidly in seed fields in late May, June, and July. These damage the flower parts and developing seed.

Various grasshopper species may move into Ladino fields in great numbers. Infestations may originate in high, dry checks and borders, and fence rows within fields. Adjacent dry fields may be breeding grounds. Yellow-striped armyworm may damage seed fields,and in some cases may strip a field in 3 or 4 days. Certain understory weeds can harbor high densities of twospotted spider mite, Tetranychus urticae Koch (Acari: Tetranychidae). This mite is mainly a secondary pest and a creature of pesticide-disrupted or stressed agroecosystems. Among the plant species suitable for use as cover crops, white clover appeared particularly prone to outbreaks of the mite. Use of herbicides led to increased movement by mites into trees.

Haley and Hogue assessed the influence of type of ground cover on apple aphid, Aphis pomi DeGeer (Homoptera: Aphididae), and its predators in a young apple orchard. Four ground cover regimes were compared: (1) fall cereal rye (Secale cereale), herbicided in spring and summer; (2) a mixture of white clover (Trifolium repens) and grass; (3) herbicided tree-row strips and grassed-in alleys; and (4) woven black-plastic strips in the tree row and grassed alleys. The trial was initiated at the beginning of the second year of the orchard.

Few aphidophagous insects of interest (e.g., the predatory mirids Deraeocoris brevis and Campylomma verbasci, the predatory midge Aphidoletes aphidimyza, lady beetles, hover flies, or lacewings) were found in the ground covers. In the first year of the study, leaf nitrogen and aphid and predator densities were lower on trees with the white clover-grass mixture. These differences did not occur the second year. Terminal growth was particularly depressed for apple trees with understories of white clover and grass.

White clover is used as a nectar source by *Scolia* sp. (Scoliidae) and *Prionyx* sp. (Sphecidae) wasps in pecan orchards of southern Georgia (Bugg, pers. comm.).

Grafton-Cardwell *et al.* found that pollen of bell bean, 'Austrian Winter' field pea, and New Zealand white clover sustained longevity and fecundity of the predatory mite Euseius tularensis (Acari: Phytoseiidae) as well as the standard diet of iceplant pollen. By contrast, reduced fecundity was observed for common vetch, woollypod vetch, and crimson clover, and E. tularensis did not survive more than one generation when fed pollen of rose clover or red clover.

Inoculation with *E. tularensis* in early spring led to build-up of the mite by late spring in a cover crop of bell bean, field pea, and woollypod vetch. Most of the mites were found on the bell bean component of the mix. When the cover crop was mowed and the mowings placed in young citrus trees, significantly increased densities of the predatory mite were observed on the citrus foliage.

Pest Effects on Nematodes

Duke listed 41 species of nematodes as having been isolated from white clover, including several species each of Meloidogyne, Heterodera, and Pratylenchus. In England, the nematode *Tylenchorynchus dubius* is known to damage white clover. In the Southeast, the root-knot nematodes, *Meloidogyne* spp. damage white clover roots. Host plant resistance, crop rotation, soil fertility, and proper cutting schedules can aid in control of these pests.

Pest Effects on Diseases

Sixty-two species of fungi have been recorded as attacking white clover, as is the bacterium Pseudomonas syringae and 14 types of virus. Miller wrote that white clover is subject to root damage by the fungi *Sclerotium rolfsii* and *Sclerotinia trifoliorum*. Root and stolon damage is caused by the fungi *Rhizoctonia, Fusarium,*

Leptodiscus, Curvularia, and *Colletotrichum.* Host plant resistance, crop rotation, soil fertility, and proper cutting schedules can aid in control of these pests. White clover can be attacked by several foliar and root diseases, as well as clover mosaic virus. The latter cause more problems where white clover persists as a perennial and regenerates vegetatively than were it reseeds as an annual.

Heavy accumulation of vegetation is deleterious to white clover because light is reduced, soil moisture may be depleted, and the microclimate predisposes for disease and insect problems. Under such a regime, white clover is prone to have very succulent growth and is unlikely to survive the microclimatic change that would result from sudden heavy mowing or grazing.

As of 1951, no serious diseases of economic importance had affected the Ladino crop in California. Interspecific hybrids between *Trifolium repens* and *T. ambiguum* show resistance to peanut stunt mosaic, clover yellow vein virus, and alfalfa mosaic virus. These resistances could prove valuable if they can be incorporated into adapted white clover varieties.

Pest Effects on Weeds

Mowing in early May of the first season can reduce weed problems; stands should be mowed before the weeds produce seed. Annual weeds such as ryegrass, burclover, canarygrass, etc., may preclude a satisfactory stand of white clover. Spring sowing of clover can circumvent this problem because the winter-germinated weeds can be destroyed during spring seedbed preparations. For seed production, grazing or mowing by May 15 can reduce weeds. Grazed fields should be mowed thereafter to control weeds and equalise the stand.

Hartl, W. reported replicated pot and field trials in which wheat was grown with undersown white clover, black medic, *Medicago lupulina* L., Persian clover, *Trifolium resupinatum* L., and various associated weeds. In both pot and field trials, biomass of weeds alive at the time of harvest was reduced substantially (50% or more reduction). The field trials showed reduction of wheat yield, as well, for black medic and Persian clover, but a non-significantly-increased yield when white clover, *Trifolium repens*, was used.

Bugg *et al.* conducted a replicated study of various cover crops (r=4) at Blue Heron Vineyard (Fetzer Vineyards), Hopland, Mendocino County, California. Weed above-ground biomass data (dry) were taken in a replicated Cover crops were seeded in late October, 1990; harvest was on May 15-16, 1991. Dominant winter annual weeds were chickweed, shepherds purse, rattail fescue, and annual ryegrass. In plots sown to a mixture of white and strawberry clovers, weed biomass was 2.8+/-0.7 Mg/ha (Mean +/- S.E.M.). This was 57.5% of the value obtained in weedy control plots. Vegetational cover afforded by the clover mix was 77.50+/-3.23 % Vegetational Cover (Mean +/- S.E.M.).

Pest Effects on Vertebrates

Some sacramento valley seed producers used night lights to protect white clover stands against ducks in the winter. Airplanes were used to herd the ducks from their fields during the daytime. Otherwise, fields may be almost denuded in a few hours.

Management Recommendations

Grazing

Large pastures of white clover should be divided and

rotationally grazed (stocked) with cow-calf pairs or with developing heifers. Small acreages can be limit-grazed and/or creep-grazed. In order to use the white clover as a protein and energy supplement, the cows can be allowed to graze only 1 to 2 hours each day. This will allow for optimum utilisation of the white clover pasture.

Irrigation

When it is cost effective, irrigation will insure maximum production and prevent loss of stand due to a late spring drought.

Reseeding

Livestock should be removed or grazing pressure reduced during the time of maximum seed production. Reduced grazing pressure should last for 2 to 3 weeks to allow some seed to mature.

Summer Live-over

To help Osceola clover plants live through the summer, the pasture grass should be harvested by mowing or grazing to a 2- to 4-inch height every 4 to 6 weeks during the summer. This will prevent the clover plants from being weakened or killed by shading.

Fall Regrowth

Clover regrowth both from stolons (summer live-over plants) and seed is again dependent on controlling grass competition. This is best done through grazing; however, mowing the grass to a 2- to 3-inch height in late September and October will be helpful if close grazing is not possible. Apply fertiliser after removing the grass. Irrigate if possible and if needed to encourage rapid development of the clover and earlier grazing.

Postscript

Occasionally someone complains that white clover just does not grow as well as it did in the 1940s and 1950s. This is probably true in many instances. It is now thought that populations of certain nematodes and viral diseases that adversely affect white clover growth have increased over the years. Also, there may be more drainage ditches than there once were, thus increasing drought stress, and also irrigation is less available than it once was. All of the factors add up to less white clover. Yet, many producers still successfully grow white clover by carefully selecting those sites that have the best chance of remaining moist throughout the fall, winter, and spring

White Clover-based Swards

Benefits of grass/clover swards compared to grass swards

- — Can produce similar dry matter yield as a grass sward receiving 200 kg of fertiliser nitrogen per hectare (6 bags of 27% nitrogen per acre)
- — Maintains high digestibility over longer period leading to improved intakes
- — Up to 10% higher liveweight gain in cattle, 20% more milk from dairy cows and 25% higher liveweight gain in sheep
- — Enhances lean meat gain and milk protein content
- — Contains more minerals, in particular magnesium, thereby reducing the risk of animal health problems associated with mineral deficiency
- — Saves on energy usage thereby indirectly reducing environmental pollution

— Evidence of reduced nitrogen loss to the environment

— Greater biodiversity

Clover is the key to grassland management. The provision of good levels of herbage production on organic units depends on achieving high clover contents in swards. The ability of clover to capture nitrogen from the air is well known but the potential value of clover-based swards is often underestimated. However, maintaining the high clover contents in swards required to achieve satisfactory levels of production needs careful sward management.

Establishing Clover-rich Swards

There are several ways of achieving clover-rich grass swards:

— Direct reseeding before the end of August is the most reliable method of establishing grass/clover swards. To ensure a good establishment of clover be generous with the clover seed and economise on the grass seed.

— The best compromise is about 25 kg of perennial ryegrass with 4-5 kg of clover seed, sown at about 29-30 kg per hectare. The seedbed should be firm and fine and should be rolled prior to sowing the grass/clover mixture.

— Undersowing can be useful for spring reseeds, using early maturing spring cereal varieties as the cover crop.

— Strip or slot seeding can be used to upgrade existing swards but the technique has given variable results under local conditions.

— A range of methods are available for placing seed into a sward, such as direct drilling using the Vertikator, very light cultivation followed by sowing with a grass drill such as the Einboch, or discing ground and broadcasting the seed.

— In all cases it is essential that grass cover is minimised by grazing the sward tightly (4-5 cm) in late July.

— Alternatively, follow in immediately after a silage cut taken mid-July or late August, when soil conditions are neither excessively wet or dry.

— Minimal cultivation represents an alternative to strip or slot seeding in that the technique does not involve ploughing. If discing, set the disc with minimum cut (coulters running straight) and add weight.

— Disc the field in several directions to open up slits in the ground before sowing when soil conditions are dry. Roll after sowing to achieve good seed to soil contact, conserve moisture and ensure rapid germination.

— Slurry seeding may be an option in circumstances where the existing sward has become thin and in need of rejuvenation.

— Pasture management to encourage clover can have a significant impact provided clover is well distributed throughout the field at the outset.

Key Factors in Reseeding

— All the basic principles for any reseeding operation must be considered as part of an overall sward improvement programme, for example, adequate drainage, pH, fertility, weed and pest control.

— Soil analysis to keep a check on soil fertility is worthwhile every 3—5 years.

— Soil samples should be taken from any fields to be reseeded. Aim to maintain a soil pH status of 6.0-6.5. Lime can be used to correct soil pH.

— In addition to manure and slurry some natural sources of phosphate and potash, such as ground rock phosphate and rock potash, can be used to maintain soil fertility.

For successful clover establishment, by any method, there are some principles that must be adhered to:

Correct timing

Late summer (August) is best for seed germination and full plant development before winter

Proper soil fertility

Soil phosphate and potash status should both be at least index 2 and soil pH should be between 6.0 and 6.5

Good seed/soil contact

Seeds need to be placed into a firm, shallow (1-2 cm) seedbed

Control grass competition

The existing grass sward cover must be kept low before and after seed sowing, otherwise the young seedlings will not survive

Encouraging clover in existing swards

— Provided clover is well distributed throughout the field, even though it may not be contributing much to sward productivity, it may be possible to encourage the development of a productive clover rich sward without the need to reseed.

- To determine the suitability of a field for rejuvenation there must be a high proportion of productive grass species present and a clover assessment should be carried out.
- As you walk across a field you must see clover within 0.5 m of your foot in 8 out of every 10 inspections, preferably carried out every 20 paces over the entire area of the field.
- If there is an even distribution of clover throughout the sward, then proceed to adopt the management guidelines given below:
- Grazing hard (3-5 cm) with sheep or light cattle during November/December
- Avoid under-grazing during spring/early summer
- Resting for 3-4 weeks during July or closing off for silage
- Avoid poaching
- Avoid smothering with slurry
- Avoid spreading silage effluent
- Control broad leaved weed development

Weed control

In direct sown swards or swards undersown in an arable silage crop, topping or forage harvesting can control many weeds. Grazing with sheep whenever the grass is 10 cm tall can provide a useful degree of control of annual weeds like chickweed, hempnettle and redshank. This can also help control ragwort in established swards.

However, care must be taken to avoid overgrazing and poaching, especially when soil conditions are wet.

Periods of frost can provide an opportunity to graze with minimal damage. Thistles do not survive long if repeatedly topped before flowering. Docks are the most difficult weed to control in an organic system.

Pest and disease control

Clover is even more susceptible to leatherjackets than grasses, although clover will eventually recover after most attacks. Although a number of diseases can affect both white and red clover there are no routine recommendations for their control.

Bloat

Bloat can occur in both beef and dairy cattle grazed on lush clover-rich swards. However, the condition can also occur on all-grass swards. As part of normal rumen function the large volumes of gas normally produced are belched and the animals suffer no discomfort. However, when bloat occurs a stable froth is produced which inhibits both normal belching of the gas and rumen contractions. Sub-clinical and clinical bloat can affect productivity, while in severe cases death can result from heart failure and from asphyxia.

Prevention

- Feed roughage, such as straw or hay, before turning out and, if necessary, during grazing
- Sheep selectively graze clover and are not prone to bloat. Grazing sheep ahead of cattle reduces the risk
- Never allow hungry cattle to gorge themselves on clover-rich pasture
- Cattle moved onto dry rather than wet pasture are at less risk

— Affected animals may be treated with anti?foaming agents

— In severe cases remove animals from clover swards and seek veterinary advice

The productivity of grass/clover swards can be more variable, both within and between seasons, when compared to grass swards dependent on inorganic nitrogen fertiliser. Clover-based swards are slow to start growing in the spring. Early growth can be encouraged through strategic use of slurry and farmyard manure. It is important that the resultant flush of grass is grazed down efficiently to allow subsequent growth and development of clover within the sward.

2

Morphology of White Clover

White clover provides a cheap source of nitrogen, has high nutritive value, improves forage intake and utilisation rates of cows, increases milk production especially in summerautumn, and complements perennial ryegrass growth patterns. White clover (Trifolium repens) is a key factor in the international competitive advantage of New Zealand pastoral agriculture which is reliant on an inexpensive, high quality feed source. In dairy pastures white clover provides a cheap source of nitrogen, (N) has high nutritive value, improves forage intake and utilisation rates of cows, and complements perennial ryegrass growth patterns. It is also considered to be environmentally friendly and therefore contributes to our "clean green" image.

Studies in Victoria, Australia and at the Taranaki Agricultural Research Station clearly demonstrated the benefits of white clover for milk production. The increased milk yield, compared with cows grazing perennial ryegrass, was due to both feed quality and quantity. White clover has a high nutritive value compared with grasses, due to lower levels of structural carbohydrate, higher metabolisable energy levels, higher digestible protein and a faster rate of passage through the rumen.

Under white clover grazing, bacterial populations in the rumen make more efficient use of proteins and amino acids. Voluntary herbage intakes (kg DM/cow/d) of cows grazing pure clover are also greater than intakes of cows grazing ryegrass. Most benefit from high clover content in dairy pastures occurs after peak lactation. The comparatively high quality of ryegrass during spring, associated with its high growth rate, makes benefits from increased clover content unlikely during spring. However in summer, clover is able to maintain its high nutritive value in contrast to ryegrass, Clover contents in dairy pastures over summer are also higher.

Optimum Clover Content for Dairy Pastures

Some researches indicated the optimum clover content of dairy pastures for maximum milk production over summer-autumn is 50-70% (% of total dry matter), not 100% clover. (Figure 1).

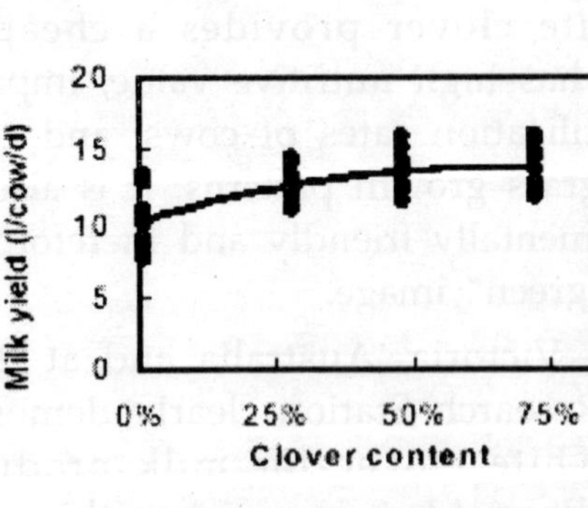

Figure 1. Effect of clover content in a perennial ryegrass - white clover pasture on milk production in February 1996.

The optimum clover content in dairy pastures for maximum pasture growth is more difficult to determine.

From an agronomic viewpoint of balancing N inputs from clover with herbage bulk as offered by grasses, a clover content of at least 30-50% has been suggested.

Average clover content at No 2 Dairy, Ruakura, during 1993-96 was 16% of total annual DM production (16.2 t/ha) on a farmlet stocked at 3.2 cows/ha and which received maintenance fertiliser but no N fertiliser. There was considerable variation in clover content both between years and at different times of the year, with levels ranging from 7% in winter-early spring, to 33% in late summer-autumn. Levels of clover on commercial dairy farms in the Waikato-Bay of Plenty are similar and show the same variation between and within seasons.

There is a general view among dairy farmers that clover content has declined over the last 15-20 years. Although there has been no long-term monitoring of clover content on dairy farms, data from various trials conducted at DRC over the past 17 years do show a decline in average clover content. Although there has been a small decline in clover content it may not be as great as suggested by visual observation. The decline in clover % has been a result of increased total pasture production because of improved ryegrass vigour and management options that favour grass growth. However, there has been no decline in actual clover yield (kg DM/ha).

High Clover Content

Despite being a desirable component of New Zealand dairy pastures, white clover levels in pastures can, however, be unreliable and unsatisfactory, such that N inputs from white clover are insufficient to obtain maximum grass growth and animal production. Mixed pastures have the ability to self regulate the N-cycle. It is this response to soil N levels which is the main factor determining ryegrass and clover contents in a mixed pasture.

Clover has the advantage at low soil N levels because it fixes N. Ryegrass has the advantage at high soil N because of its more efficient uptake of mineral N and greater competitive ability for carbon fixation. During periods of clover growth, clover increases soil N levels and stimulates grass growth. However, during periods of grass dominance, grass will draw down soil N levels. This cycle of grassclover growth makes it difficult to increase the clover content since exploitation of clover N fixation by grass leads to at least as much increase in ryegrass levels as in clover levels. This cycle, together with numerous other factors affecting clover growth, makes it very difficult to establish, and particularly maintain, high clover contents in pastures without resorting to impractical and uneconomic managements.

It is possible to increase clover levels to at least 50% in the shortterm, e.g. by establishing new pasture after several years of maize using a high clover seed rate (5 kg/ha). However, it is very difficult to maintain a high clover content in a mixed pasture since clover levels decrease over time and a ryegrass-clover balance is reached. Dairy farmers can, however, maintain clover levels and possibly achieve small increases in clover content by helping to increase clover's competitive ability.

Clover Content in Dairy Pastures

A number of factors affect clover content in dairy pastures. Addressing these factors while still maintaining a profitable farm system is the key to maintaining or achieving small increases in clover levels in pastures.

Nitrogen

Maximum gross margins per ha and a high level of milksolids production per ha and per cow are best achieved by combining N inputs from white clover and N fertiliser. Clover content can be maintained at 15-20% even when applying up to 200 kg N/ha/y, provided the pasture grown is fully utilised, particularly in spring. Applying the N as split dressings of less than 50 kg N/ha is recommended so that the effect on clover is reduced.

Soil fertility

White clover is a poor competitor for phosphate, sulphate and potassium, compared with grasses. Without adequate levels of P, K and S, white clover yield and persistence is jeopardised. The ideal levels for intensive dairying are 25-30 Olsen P, 8-10 K and 8-10 S. Lime is important for successful establishment of clover on soils where pH is below optimum (pH 5.5). Lime also promotes nodulation, better use of P, and increased molybdenum availability which is essential for N fixation in clover.

Companion species

Grass species have different competitive abilities with white clover which can influence clover content of pasture. Perennial ryegrass and tall fescue are more compatible than cocksfoot, Yorkshire fog, and many of the subtropical grasses such as kikuyu. Even with ryegrass it has been demonstrated that when tiller numbers exceed $5000/m^2$, white clover content is decreased. N fertiliser can increase ryegrass tiller density and therefore reduce clover content.

Clover cultivar choice

There is now a range of white clover cultivars available to

New Zealand dairy farmers for use in pasture renovation. Larger leaved types such as Grasslands Kopu, Aran, Grasslands Challenge, Grasslands Sustain, Le Bons and Grasslands Pitau, should result in higher clover contents than medium and small leaved types. In a four year white clover cultivar trial at DRC, Challenge, Pitau, Sustain and a breeding line, G. 49 x Crau, yielded the best in mixed pasture. However, neither Challenge nor Pitau persisted well in the fourth year. Cultivars with excellent winter growth include Challenge, Kopu and Pitau, and those with good summer growth include Aran and Sustain.

Clover establishment

Care taken during establishment of new pastures will be rewarded with a high clover content for at least the first two years after sowing. Full cultivation provides the best seedbed for clover establishment, but at a higher cost than drilling clover into permanent pasture. Weeds should be removed before renewal by either spraying before drilling, or by spraying before cultivation. Soil drainage or fertility problems should be remedied before final cultivation, while cultivation will help reduce soil compaction problems. A level, fine, firm seed bed should be prepared if cultivating.

Seed is best sown using a 'V-ring' roller drill and lightly covered with chain harrows and then rolling again (except on strong clay soils prone to caking). Clover seed should be sown no deeper than 10-15mm—this is particularly difficult if drilling clover seed into existing pasture. Burial of clover seed at too great a depth, especially when clover is sown in combination with ryegrass, is the main cause of poor clover establishment. A higher clover content can be achieved by using a higher seed rate, e.g. 5 kg/ ha instead of 3 kg/ha.

However, the resulting increased clover content is short-lived, making the profitability of this questionable. It is not known whether inoculation of clover seed improves performance or persistence. It is best, therefore, to use bare seed or to inoculate within 12 hours of sowing. Annual broadcasting of additional clover seed into established pasture, often in combination with autumn fertiliser application, may sometimes boost clover content.

Grazing management

Attaining a stable composition of white clover in pasture is affected by the preference grazing animals, including dairy cows, show for clover. When offered both pure clover and pure ryegrass, animals prefer a mixed diet of about 70% white clover : 30% ryegrass. However, this level of preference is unlikely to be achieved by cows grazing a typical mixed pasture since selection is largely restricted to clover patches in the sward which have a higher clover content than the pasture average, rather than at the level of fine discrimination between interspersed pasture species.

While selection of white clover by grazing cows may reduce clover content, it is also desirable so that the nutritional benefits of clover can be maximised. Although studies have shown total herbage production from grass/clover swards increased as defoliation interval was lengthened, overall clover content was not generally increased. At certain times of the year, however, shortening or lengthening grazing interval may increase clover content.

For example, L'Huillier showed rotation length over winter had no effect on clover content, but short spring rotations (8 days) resulted in higher clover contents (36%

vs 18%) than longer rotations (32 days). Ensuring ryegrass growth is kept under control during spring, particularly if N fertiliser is used, may not increase clover content but will at least maintain it. Clover content in pastures following a silage cut, particularly after close cutting (3-4 cm), is generally high, despite herbage levels being very high prior to cutting. This is because clover is able to adapt to paddocks being closed for silage in spring by forming longer petioles and avoiding shading, provided paddocks are not closed up for too long. Trials at DRC have also shown that lengthening the defoliation interval, or using deferred grazing during dry summers, increased clover content. Grazing down to a low residual (<1400 kg DM/ha) in summer may lead to high soil surface temperatures, resulting in death of clover plants.

Pests and diseases

Many pests and diseases reduce both yield and persistence of white clover in dairy pastures. These include grass grubs, root invading nematodes, porina caterpillar, clover flea, slugs, fungi such as rust, and the white clover mosaic virus. Root knot (Meloidogyne spp.) and clover cyst (Heterodera trifolli) nematodes reduce clover yields by 30% and nitrogen fixation by up to 58% with the resulting poor clover performance likely to decrease milk production. Unfortunately, nematodes occur in all soil and while pesticides can control them, it is not economic. Breeding resistance, using genetransfer technologies, or the introduction of biological control agents, may be the most practical solutions to pest and disease problems, although this does not provide a short-term solution for dairy farmers.

Recently, the clover root weevil (*Sitona lepidus*) has appeared in New Zealand and is potentially a serious pest of white clover. Since its accidental introduction three years ago, the weevil has spread from Waipu Cove in the north to Te Kuiti in the south and Opotiki in the east. The weevil attacks only clovers and shows a preference for white clover over other clover species. Its impact on white clover yields is still unknown. Small weevil larvae feed on root nodules, larger larvae feed on roots, while the adult weevils feed on the foliage.

On some dairy farms the weevil has been blamed for the loss of white clover from pastures. Both the larvae and adult weevils are present throughout the year, making management of this pest difficult. As yet there are no known pesticides or biological control agents for use against the weevil. However, healthy clover plants will be more tolerant of insect attack.

Soil moisture content

Low soil moisture levels, particularly in conjunction with high temperatures (>30°C), can have a detrimental effect on clover content. White clover has less drought tolerance than many grass species, due to a shallower root system and poorer water use efficiency. Selection of white clovers with a deeper root system through increased taprootedness has shown some benefit in dryland environments. Clover content of pastures can be maintained or increased during dry periods by irrigation, as demonstrated during dry summers in Northland.

Grazing Condition in White Clover

Grazing increases the importance of genotype, management and edaphic and climatic conditions in determining white clover persistence. For example in Brittany interaction between soil and climate results in clover persistence being poorer in wet than in well drained soils on farms due to increased risk of treading damage.

Components of Grazing

This chapter considers the effects of the components of grazing on growth and development of white clover and their relative importance in determining persistence of white clover.

Defoliation

Generally white clover responds to defoliation by producing leaves with shorter petioles and smaller laminae and shorter internodes than in undefoliated plants. Small leaved varieties are less productive under long defoliation intervals. While leaf appearance rates are relatively insensitive to cutting interval unless it is extremely short or long, stolen internode length and diameter increase with defoliation interval.

Severe defoliation of clover plants, ie rapidly repeated low defoliation, reduces branching, presumably due to limitation in the availability of assimilate; so axillary buds being weaker sinks than the apical meristem are inhibited from developing. Incorporating a longer rest period among weekly defoliation intervals at 3.5 cm has improved leaf appearance and branching rates in clover and reduced them in grass when the weekly defoliation intervals have resumed, compared to the control without a rest period.

Relative rate of removal of clover and grass by grazing is an important facet of defoliation. The complexity of selective grazing of clover in a grass/clover sward has been demonstrated by Parsons *et al.* Under hard, continuous grazing, even if an animal has a strong preference for clover, eventually the composition of the diet will reflect the relative rates of growth of the two components. Therefore rate of clover removal by grazing cannot be considered in terms of one co-efficient for selection or preference (especially for sheep continuously grazing).

Treading damage

Treading damage is a consequence of stocking rate, soil conditions and season. Relative to perennial ryegrass white clover is slightly less tolerant to treading although the difference is dependent on stocking rate. In addition to leaf production per unit area stolon length per unit area is affected, the depression again being related to stocking rate. While this effect may be due, in part, to direct damage burial of stolon by grazing animals (especially cattle), tramping stolons into soft ground can also influence stolon development. In New Zealand as a consequence of earthworm casts and treading Hay *et al.* have recorded about 80% of total stolon weight to be below the ground surface level in mid winter in sheep-grazed plots.

The impact of burying stolons under a shallow layer of soil, with or without defoliation in a pot experiment has been assessed by Grant *et al.* They found that conditional on the stolons not being defoliated, stolon extension, leaf appearance and water-soluble carbohydrate concentration in the stolons remained unaffected. However, distribution of axillary buds and inflorescences differed to that in the unburied controls, a higher proportion being on the new aerial segments of stolon but less on the secondary stolons. By combining defoliation (removal of all expanded leaves weekly) and burial almost half of the secondary branches died after 1½ months.

Fewer axillary buds on secondary stolons developed and although a higher number developed on new segments of primary stolons these were insufficient to compensate. Clover can recover quickly from autumn treading after a mild winter, stolon fragments developing into separate plants and earthworm activity encourages adventitious root development. Grazing in a wet winter may hinder earthworm activity, delaying recovery of soil structure. Damage due to treading in spring may be greater than in autumn as grass recovers quicker than clover and so competes vigorously with clover.

Faeces and urine

Even in the absence of N fertiliser, N returned in faeces and urine to a grass/clover sward may be considerable. Table 1 shows that more N is returned in dung and urine to a grass/clover sward fixing 120 kg N/ha than to a grass sward receiving 250 kg N/ha.

Table 1. Nitrogen returned annually in faeces and urine in a grass and grass/white clover sward grazed with heifers in Brittany (1991).

	Ryegrass	Ryegrass/white clover
N applied (kg/ha)	250	
N fixed (kg/ha)		120
Faeces (kg/ha)	6900	6200
N%	0.32	0.46
N returned (kg/ha)	22	28
Urine (l/ha)	13500	18500
N%	0.70	0.79
N returned (kg/ha)	94	147

The direct effect of dung and urine on clover growth and development has usually not been separated from the effects due to stimulated grass growth. However, Curll and Wilkins interpreted a reduction in clover content in grass/clover swards due to excretal returns to be caused mostly by stimulated grass growth, stolon length per unit area being affected more than clover growth. In contrast, applying urine to grass/clover, Marriott *et al.* found that while reduction in clover growth and branch number was considerable the stimulatory effect on grass growth was modest. Under cattle grazing, urine also reduces N2-fixation but increases both grass and white clover growth. However, scorching (during dry periods) and loss of N by leaching (during wet periods in summer and towards the end of the growing season) modify this general effect.

An additional effect of faecės return on grazed grass/ clover swards is the rejection of herbage around dung pats resulting in changes in grass and clover morphology. In mixed swards continuously stocked with cattle stolen elongation rate was faster but branching rate slower in rejected compared to grazed areas the major effect being due to shading by the accumulating ungrazed herbage. The patches of bare ground resulting from dung smothering the sward are potential sites for clover stolons to colonise, although the relatively high fertility of the area may be a constraint.

Grazing Management

The impact of the above components is influenced by type of animal, grazing system and grazing intensity and so system and intensity can be manipulated to some extent to control clover development. Generally, growing point density will increase with stocking rate, within limits, in swards continuously grazed by sheep or cattle. While this might seem to contradict the point that clover branching is reduced by defoliation it can be explained by the adverse effects of shading on branching by the accumulating herbage in the tall swards (death or inhibition of secondary axillary buds, lower leaf appearance rates and leaf weight ratios).

Some cultivars of clover are able to avoid continuous defoliation by reducing the size of new leaves e.g. Kent wild white compared with Huia and so continue branching and aid persistence. Comparison between grazing systems generally shows continuously stocked swards to have a higher stolen population density than rotationally grazed swards but there are exceptions. Different defoliation patterns between grazing systems obviously influence the morphology of clover, regrowth allowing larger leaves and more leaves per stolon to develop.

Manipulation of grazing intensity or system at strategic times can improve white clover stolon density eg. grazing to a low height in late autumn or early spring or continuous stocking in spring followed by rotational grazing throughout the remainder of the growing season or interrupting continuous stocking in summer with a silage cut.

Out of season grazing also increases white clover potential photosynthesis presumably due to lower competition from accompanying grass. Interrupting continuous stocking with a silage cut in summer results in a high proportion of both grass and clover leaf being removed but the laminae in new clover leaves remain large when grazing resumes conferring a competitive advantage on clover. While success of clover in a sward is often assessed in terms of stolon population density, a maximum can be reached beyond which total plant size and complexity declines followed by a marked decline in plant number.

3

Root Habits of Corn Maize

Corn or Maize (*Zea mays ssp. mays*) is a cereal grain. Corn (*Zea mays indentata*) is a summer annual. It has a coarse fibrous root system which spreads widely and penetrates deeply. The depth of planting apparently bears no relation to the depth of rooting. The first whorl of roots usually arises within an inch or two of the soil surface, even if the seed is planted several inches deep. In this respect, it is similar to other cereals.

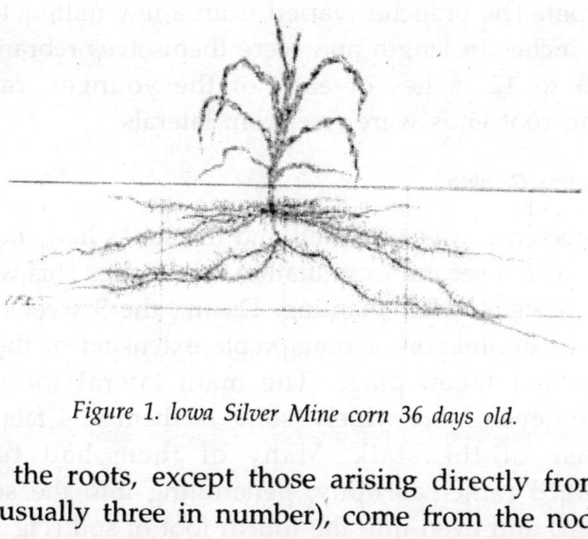

Figure 1. Iowa Silver Mine corn 36 days old.

All of the roots, except those arising directly from the seed (usually three in number), come from the nodes in whorls. But since the internodes are very short, careful examination-is often necessary to ascertain the origin of the 2 to 10 or more roots from each node. The entire group of whorls constitutes the root crown.

Root Development

The general root habit is well illustrated by Iowa Silver Mine corn grown in loess soil in eastern Nebraska. This crop was planted on May 9, after the soil had been shallowly plowed and harrowed. The rows were 3 feet apart, and the kernels were drilled 3.5 inches deep and 1.4 feet apart. When the corn was, 5 weeks old, it had the root habit shown in Fig. 1. Both the seminal and adventitious roots were confined almost entirely to the surface foot of soil. The roots were coarse, from 10 to 15 in number, about 1.5 millimeters thick, and ranged in length from 1 inch to over 2.5 feet. Throughout their length, except near the tip, they were profusely branched, 33 rootlets sometimes occurring on a single inch of the main root. The branches varied from a few millimeters to over 4 inches in length and were themselves rebranched. Only 8 to 12 inches of each of the younger, rapidly growing root ends were free from laterals.

Midsummer Growth

When the crop was 4 feet tall and the stalks had about 12 leaves each, a second examination was made. This was on July 5, 8 weeks after planting. During the 3 weeks since the first examination, a remarkable extension of the root system had taken place. The main lateral roots had extended even more widely, some of them to 4 feet from the base of the stalk. Many of them had turned downward rather abruptly, penetrating into the second and third and even into the fourth foot of soil (Fig. 2). In addition to these, an entirely new group of roots had arisen. These penetrated more or less vertically downward 3 to 4 feet and filled the soil beneath the plant. This region had not been penetrated by the main lateral root system. Thus, the main vertical roots supplemented the main lateral ones. The longest of the vertically penetrating roots had grown at the rate of over 2 inches per day.

That growth was not yet complete was shown by the abundance of new roots, some of which were only a few inches long and had succulent turgid ends 3 to 4 millimeters thick. Moreover, it was clear, from the absence of root hairs and branches near the ends of even the oldest roots, that they were still elongating. The longest branches were confined to the first foot of soil. Many extended to within half an inch of the surface. These, with their many ramifications, formed an intricately dense and efficient absorbing system.

At this time, the roots of neighbouring stalks, which were only 16 inches apart, had greatly overlapped. Indeed, roots of plants in adjoining rows, 3 feet apart, were drawing upon the same soil area for water and nutrients. Since many roots grew so near the surface, the field was cultivated shallowly with a hoe in order not to disturb them. The large quantities of water transpired by corn in eastern Nebraska have been measured, 275 pounds or 34 gallons (over a barrel of water) usually being needed to mature a single stalk of corn.

Mature Root System

The root system of maturing plants was excavated and examined on Sept. 2. The stalks were 8 to 9 feet high, and though a few leaves had dried most of them were still green. The husks on the ears were beginning to dry, and

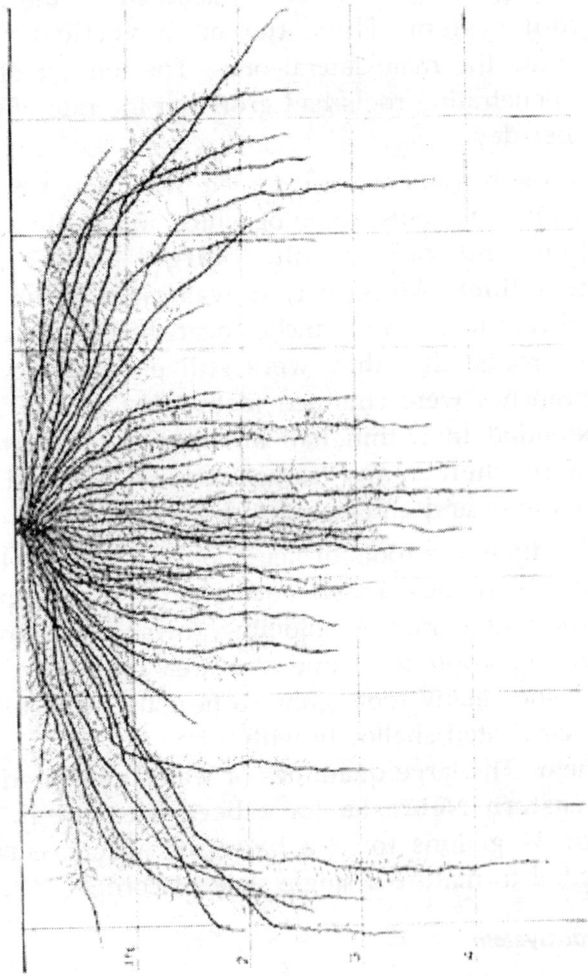

Figure 2. Root system of corn 8 weeks old.

the kernels were dented. The crop had completed its root growth.

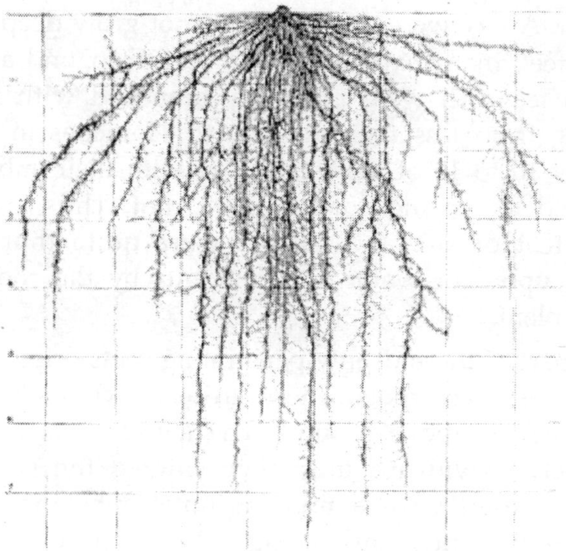

Figure 3. Mature root system of corn.

The main lateral root system (Fig. 3) had scarcely increased over that found early in July. Most of the roots of this type were found in the first foot or two of soil. Some retained a nearly horizontal position throughout their entire course. Others ran at various angles from a few inches to 3 feet or more and then turned downward either abruptly or with a gentle curve. The longest extended into the third and fourth foot of soil, or even deeper. Branching throughout was even more profuse than before. Unlike the shallower portion of the root system, the more deeply penetrating part had made a marked development.

The number of roots varied from 20 to 35. They either ran straight downward from the base of the plant or obliquely outward to a distance of 2 feet or more and then, with a graceful curve, took the perpendicular line of growth. A few were short and did not grow deeper than 1 to 2 feet, many extended 6 to 7 feet deep, and a few to over 8 feet. All were profusely branched with laterals ranging from less than an inch to 15 inches in length. Usually 10 to 12 of these branches with their sublaterals occurred on an inch of the main root. Thus, over 200 cubic feet of soil and subsoil were quite thoroughly drawn upon for water and nutrients by the roots of a single plant.

A part of these deeply penetrating roots originated as. "prop" or brace roots. These arose in whorls from the lower nodes above ground. The aerial roots extended obliquely downward until they entered the soil. They were covered with a mucilaginous substance which protected them from drying. The portion in the air stratum was unbranched, but after entering the soil, they could not be distinguished from the other roots. They performed the double rôle of anchorage and absorption.

Summarising, the shallower part of the root system (main lateral roots) completed its development early, so far as lateral spread was concerned. The main vertical roots developed later and continued to increase in number and extent until well towards the time of maturity. The lateral spread on all sides of the plant was approximately 4 feet and the maximum penetration 8 feet. Thus, corn is furnished with a remarkably extensive and efficient root system.

Root Habits

A study of the root habit shows clearly why corn does

best on a deep, well-drained soil which has an abundant and uniform supply of water throughout the growing season. If the soil is well prepared before planting, the main benefits of cultivation are derived from keeping own weeds, preventing the crusting of the surface, and keeping the soil receptive to rainfall. The superficial position of the roots shows clearly why deep cultivation is harmful. Fortunately, weeds are most easily destroyed when coming through the surface of the soil by shallow cultivation such as harrowing surface-planted corn. This also breaks the soil crust, giving a drier and warmer soil and more vigorous crop growth. The harmful effect of letting weeds grow for a time is not entirely due to their rapid removal of water and plant food materials from the soil, but to the breaking of the corn roots due to the deeper cultivation necessary for weed eradication.

An examination of the half-grown root system explains why late tillage, except for weed eradication, is of little value. The roots are so well distributed through the soil that little moisture can escape even from uncultivated land. Hilling at the last cultivation not only acts as a mechanical support to the stem but also encourages the development of brace roots which are an additional aid in holding up the plant against strong winds. But if hilled early, later cultivation partly removes the hill and exposes a portion of the root system. Even shallow cultivation cuts off many of the roots, and deep cultivation is very harmful and greatly decreases the yield.

Cultivation to a depth of 4 inches during a period of 9 years in Ohio gave a decreased yield every season but one, as compared with similar cultivation to a depth of 1.5 inches. The average decrease per acre was 4 bushels of grain and 183 pounds of fodder. In Indiana, very similar results have been obtained. In Missouri, deep cultivation compared with shallow reduced the yield 6.5 to 13 bushels per acre. The harmful effects of deep cultivation are always more pronounced during years of drought. In Illinois where the roots were pruned to a depth of 4 inches at a distance of 6 inches from the hill, the yield, was decreased 17 bushels per acre.

For the highest yields, cultivation should never be deep enough to injure the roots seriously. They should be allowed to occupy fully the richest portion of the soil, which is usually the furrow slice. The proper type of cultivation is one which is deep enough to kill the weeds but shallow enough to reduce root injury to a minimum. Why listed corn stands dry weather better and why this method of growing corn is more successful in regions of limited rainfall and on comparatively light types of soil may be explained in part by root development.

The root system begins its growth deeper in the soil, is further covered by each tillage, and is, therefore, not so subject to drought as is that of surface-planted corn. Listed corn stands up better and is very rarely blown down on account of the roots pulling out. This is probably due in part to the more favourable conditions stimulating the growth of prop roots.

The wide spread of corn roots is an important factor in competition. Where the soil will stand a heavy rate of planting, as moist bottom land, undoubtedly the increased yield of drilled over hill-planted corn is due in part to the better distribution of the root system. Another reason, aside from better light, may be due to the fact that the roots are damaged less than under the hill method where cross cultivation is practiced.

It has been clearly demonstrated that the results obtained by the "ear to the row" method in variety selections, etc., are often vitiated by the competition of roots and tops of plants growing in close proximity. Why the rate of planting should be decreased in drier and poorer soils may be readily understood by examining the extensive root system and the large amount of water absorbed by the corn plant.

Stages of Root Development

Like most other plants, corn shows marked variations in root habit when grown in soils of different texture and chemical composition, since these not only affect available nutrients but water content and aeration. Some illuminating experiments were conducted at Greeley, Colo. Here, in fine sandy loam soils of very similar physical and chemical composition, crops were grown under different amounts of, irrigation. The annual rainfall is only 13 inches and irrigation is widely practiced. A yellow dent corn, Minnesota No. 13, a variety considered the best adapted to this region, was used. It was grown in plots which were all treated alike as regards seed-bed preparation, time and rate of seeding, shallow cultivation, etc. One of the plots received no irrigation water, another was irrigated only lightly, and the third was fully irrigated. Moreover, the irrigated plots had been fertilised uniformly with 5 tons of barnyard manure per acre.

Root system of 6-weeks

At the end of 6 weeks, when the crop was 12 to 15 inches tall, the root systems were examined. Marked differences were apparent. In the fully irrigated soil, where both water and nutrients were plentiful, the roots, as is normally the case, were almost all in the surface foot. Nearly all were quite parallel with the soil surface, and in fact, the bulk of them were in the surface 6 inches of soil. Only a few ran obliquely downward and entered the second-foot layer (Fig. 4). But in the dry land, where only 4 to 5 per cent of available water was present, the direction of growth was different.

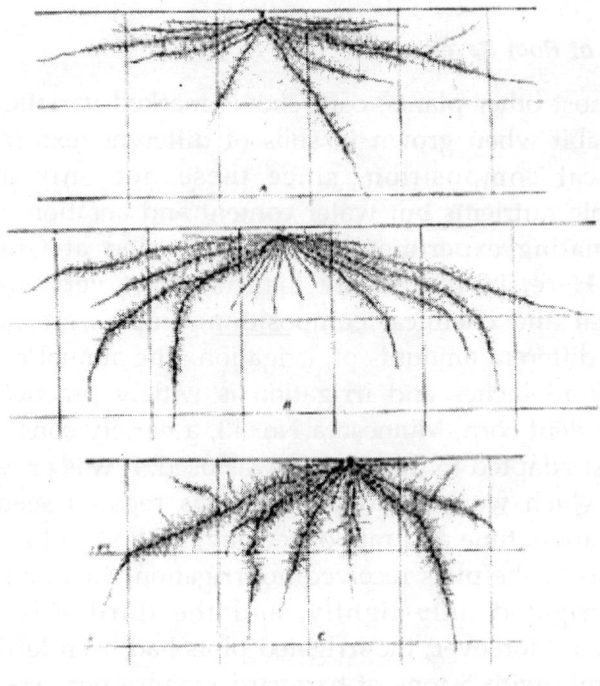

Figure 4. Root system of 6-weeks-old corn: A, fully irrigated soil; B, lightly irrigated; and C, dry land.

Although a few roots ran rather horizontally, most of them grew outward and downward 1 to 2 feet. The longest extended 2.5 feet, where they turned downward into the second foot of soil. In fact, this latter tendency was marked, roots seeming to seek moister areas. Some of the more nearly vertical roots even extended a little into the third foot. Root habit in the lightly irrigated plot was intermediate. Although the number of roots was the same in all three plots, there were great differences in both number and length of branches. In the fully irrigated soil, they averaged 12 per inch of main root as compared with in dry land. Branches in the fully irrigated soil were only 2 inches long, with 9 branchlets per inch on an average; in dry land, they averaged twice as long and also had twice as many sublaterals.

Root system on midsummer

Further studies on July 11, 17 days later, showed even greater differences. No efficient rain had fallen, and only 2 to 4 per cent of available, moisture occurred at any depth in the dry land. Water in the surface 6 inches was entirely exhausted. The corn was beginning to tassel in all the plots, although it was only 2 feet tall in the dry land, where it showed the effects of continued drought. The leaves were tightly rolled during the hottest part of the day, and sufficient water was not available for them to recover even during the cool nights.

In response to the low water content, the roots had made a more extensive growth in the dry land than in either of the other plots. The lateral spread had been increased to a maximum of 42 inches, furnishing water from a region hitherto unoccupied. This exceeded that of the fully irrigated plants by 14 inches. The vertically penetrating main roots were well developed, reaching a working depth of 30 inches and a maximum penetration of 46 inches. This was very different from those in the moister manured soil (Fig. 5). Here, although the stalks were a foot taller, three-fourths of the root system was still limited to the surface foot. The working level was only 24 inches and the maximum penetration 4 inches greater. Branching, as at the preceding examination, was much less pronounced and the laterals were very much shorter. The soil of intermediate water and air content contained roots that were intermediate in many respects to those described.

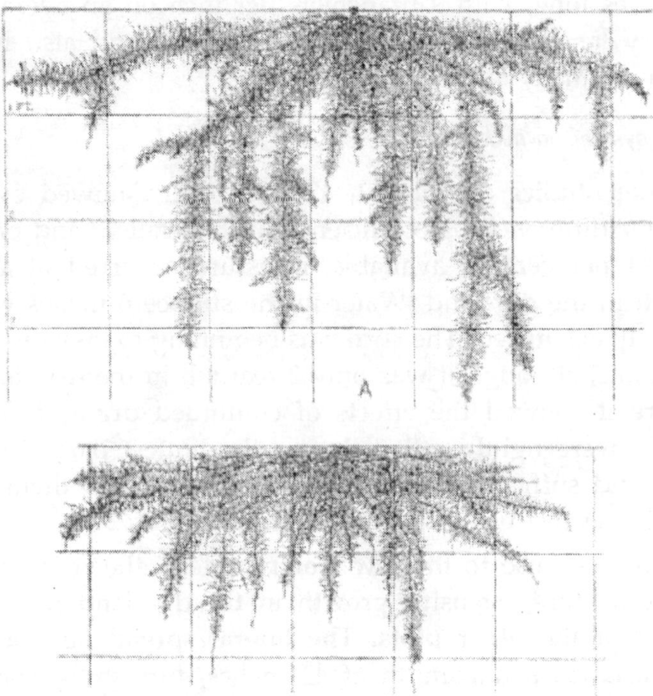

Figure 5. Root system of corn about 8 weeks old: A, dry land; B, fully irrigated soil.

Mature root

During September, the mature plants were examined. In the dry land, drought was so severe that the corn had "fired" and lost most of its leaves during July. It was only 3.5 feet tall, and the ears were very poorly developed. The irrigated corn was 7 feet tall and indicated a heavy yield.

Practically no change had occurred in the roots in dry land, except a more thorough occupation of the soil directly below the plant. Branches now occurred on all the roots to their very tips.

The roots were unable to furnish sufficient water to keep the leaves turgid and these in turn failed to supply the materials needed for further root growth. But in the fully irrigated plot, marked changes had occurred. The roots, formerly confined to the first foot or two of soil, now extended into the fifth and sixth foot, and many of the horizontal ones had likewise turned downward and penetrated deeply. A working depth of 40 inches and a maximum penetration of nearly 6 feet were found. Seepage water occurred at the 6- foot level. This fine root system was surpassed, however, by that in the drier and somewhat more sandy soil of the lightly irrigated plot. As in the former case, there were 50 to 70 major roots. The root habit was practically identical with that grown in loess soil (Fig. 3), the working depth and maximum penetration being about a foot less.

It is interesting to note that the yield was in direct relation to the extent of the mature root system. This was at the rate of 25 bushels per acre in dry land, 102 bushels in fully irrigated (later, water-logged) soil, and 115 bushels in the lightly irrigated plot.

Root development under increased precipitation

Owing to the better watered soil, the young roots showed a more normal surface spread with less tendency to turn downward. The branches were also fewer and shorter than during the preceding year. The top growth was far more luxuriant (yield, 51 bushels per acre) and the roots too were much more extensive. They had no greater lateral spread, but the working level (about 4 feet) was 16 inches deeper and the maximum penetration was 5.5 feet. The length of branches was again greater than in the fully irrigated plot where root behaviour was almost identical with that of the preceding year.

Moreover, owing to smaller differences in water content than formerly, root habit in the lightly watered plot was more nearly like that in the fully irrigated one. These results show clearly the profound modifications that are brought about by differences in environment, Since, in these experiments, the aerial environment, i.e., temperature, humidity, wind, Aevaporation, etc., was nearly the same in the several plots, the differences were due to edaphic or soil factors. Among these, as has been pointed out, physical and chemical composition of soil was very much the same (except for manuring), and temperatures at all corresponding depths were almost identical. Hence, the controlling factor in root variation was water content.

The root habits of corn have been examined more or less thoroughly, usually in connection with tillage experiments, in widely separated areas in the United States. At Geneva, N. Y., the lateral spread was found to be 3.7 feet, 14 but the roots were traced to a depth of only 2.8 feet. In stiff clay soil in Pennsylvania, corn seemed to be a shallow-rooted plant. In Illinois , root habits have been observed which were very similar to those described, the plants having a depth of 6 feet or more. At Madison , Wis., it has been found that corn roots grow 4 feet deep and in well-drained soils even deeper, under good cultivation drawing upon soil water in considerable quantities at depths greater than 7 feet. Corn roots commonly reach a depth of 3.5 to 4 feet at Fargo, N. D., and at Manhattan, Kan., they have been found at depths of 4 to 5 feet. Corn grown at Garden City, Kan., in sandy loam soil which was irrigated in the fall, after plowing, with 8 to 10 inches of water, had much deeper roots.

A lateral spread of 3.7 feet and a depth of 6 feet were found. Root development at various stages of growth were in all essentials similar to those described.

In comparing these results, it should be kept in mind that disagreements are due, in part, to environment and also to varietal differences in the crops examined. Moreover, in some cases, the roots were not traced to their extremities. Experimental evidence indicates that there are fundamental differences in root systems of various inbred strains of corn. Reduced root systems occur among strains susceptible to root rot and leaf firing. Not only do they have significantly smaller numbers of main roots than plants of good strains, but the number and length of lateral branches are also much less.

In some cases, there seems to be an actual deficiency in the root system compared with the vegetative growth above ground. Certain strains have such limited and inefficient root systems that they are unable to function normally during July and August, when soil moisture is low. Plants with reduced root systems are much more susceptible to lodging and give a lower grain production.

4

Root Habits of Sorghum

Sorghum is a genus of about 20 species of grasses raised for grain, native to tropical and subtropical regions of Eastern Africa, with one species native to Mexico.
Sorghum is a genus of about 20 species of grasses raised for grain, native to tropical and subtropical regions of Eastern Africa, with one species native to Mexico.

Figure 1. Sorghum

All of the many varieties of sorghum (*Andropogon sorghum*) are annuals. They have a well-developed root system similar to that of corn but generally finer and more fibrous. Shortly after germination and the growth of the radicle or seed root, 14 the first node develops below the surface and from this node the first permanent roots develop. After the permanent roots begin to function, the temporary root [radicle] soon decays.

Black Amber Sorghum

The Black Amber variety of the sorgo group was grown in both upland and lowland soil at Lincoln. Three months after sowing, when the plants were full grown and the seeds were maturing, the root habit was studied. The tough fibrous roots were 3 to 4 millimeters in diameter at their origin and often 1 millimeter thick at a depth of 4 feet. They were of a grayish-white color, those near the surface being strongly tinged with red, but the deeper, younger roots were glistening white. They were very numerous and completely occupied the hard, dry soil beneath the plants.

Branches from ½ inch to 3 inches long were exceedingly abundant and were quite well rebranched. In the more mellow surface soil, these branches spread somewhat widely in all directions, but in the hard, jointed subsoil, the branching was confined largely to the crevices and was in one plane. The abundant ultimate branches were hair-like, shining white, and exceedingly delicate. Not infrequently, they occurred in clusters of three to five on a millimeter of root length. Often, they formed cobweb-like mats covered with root hairs in the deeper soil crevices.

As a whole, the absorbing system of sorghum is very efficient and so well distributed throughout the soil that it can thoroughly exhaust it of the water available for plant growth. In both plots, the roots reached a working depth of 4 feet, although the average height of tops, 4.4 feet, of the more vigorous plants in the lowland exceeded that in the upland by 1.4 feet. A maximum depth of about 4.6 feet was attained.

Folger Sorghum

The Folger variety of sorgo, was examined in a compact, upland, sandy loam soil at Manhattan, Kan. Plants 2.5 months old had roots which resembled those of corn, although the fibrous growth of roots near the surface was much less prominent. Midway between the hills, which were 3.5 feet apart, the roots were 6 inches below the surface, and at the hill, the average depth was nearly 3 inches, which was fully as deep as the seeds were planted. Some roots reached depths of 3 feet. At maturity, the surface soil was much more thoroughly occupied, often to the ground line; the working depth was also greater, some roots extending to the 3.5-foot level.

Kafir

Kafir corn of the Blackhulled white variety, grown at the same Station, had roots penetrating to a depth of nearly 3 feet when the plants were 4 feet tall and nearly 2.5 months old. But the greater number did not penetrate below 18 inches. The roots, though similar to those of corn, were finer and more fibrous, but coarser than those of the sorgo roots just described. The vertical growth was much less strongly developed than in corn, but roots in the surface soil were more strongly developed. At maturity, the greater portion of the roots did not exceed 3 feet in depth, although some were found at 3.5 feet.

Blackhull Kafir and Dwarf Milo

The root habits of the grain sorghums, Blackhull kafir and

Dwarf milo, have been thoroughly studied in sandy loam soil at Garden City, Kan. The crops were grown in rows 3.7 feet apart, the plants being, 8 to 18 inches apart in the row. The plots were well irrigated the preceding autumn but not cultivated after planting, the weeds being kept out by scraping the surface with. a hoe. Roots were examined at four stages of development. When the tops were 1 foot tall and in the 6- to 8-leaf stage, 12 to 15 roots from each plant penetrated to a depth of 1 foot, although some of them were 1.5 feet deep. The lateral spread was 3 feet and the upper soil was well filled with roots to within ¼ inch of the surface (Fig. 2).

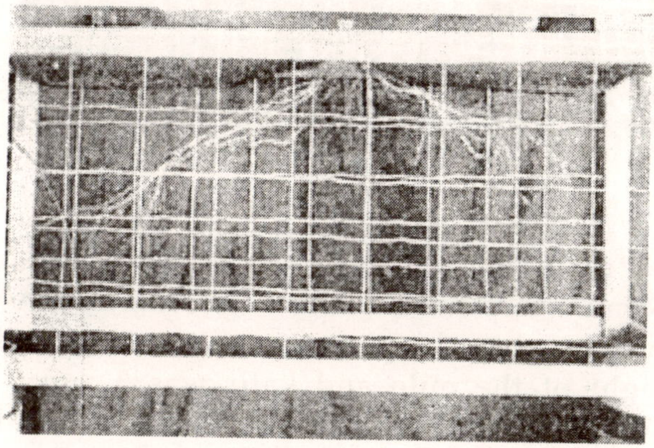

Figure 2. Root system of a Dwarf milo plant on June 24 at the age of 4 weeks

Maximum lateral spread is 3 feet. When the plants were 7 weeks old and 2.5 feet tall, the wide lateral: spread of nearly all of the roots was very marked. Although they reached out laterally 3 feet in milo, yet none were over 2.8 feet deep. The kafir was only 2.5 feet deep but had a lateral spread of 4 feet. Branching was profuse throughout, an average of 30 laterals occurring per inch of main root.

The lateral spread had increased to about 3.5 feet for milo, and both varieties had greatly increased their absorbing area in the deeper soil.

Figure 3. Root systems of two Dwarf milo plants On Sept. 3 when the seed was in the milk stage.

A maximum depth of 4 feet was attained. A final examination, when the plants were maturing seed and the height of the milo and kafir was 3 and 5 feet, respectively, showed still further root growth. Both had a very profusely branched root system reaching 3 to 4 feet on all sides of the plant and penetrating to a depth of 6 feet (Fig. 3).

Drought Resistance and Growth in Poor Soil

The fundamental characteristics possessed by the sorghums which enable them to withstand severe climatic conditions better than corn. In these experiments, Pride of Saline dent corn was grown in alternate rows with the sorghums. It is worthy of note that the primary and secondary roots of both sorghums at all stages of growth were more fibrous than those of corn. Each of the three kinds of plants at any period of growth possessed the same number of primary roots, and the general extent of these roots in both a horizontal and a vertical direction was the same for all.

The length of the secondary roots was also found to be approximately the same at any examination, but it was ascertained that the number of secondary roots per unit of length of primary root was approximately twice as great for the two sorghums as for corn. Moreover, this root system, which, judging from the number of secondary roots, would be twice as efficient in the absorption of water, supplied a leaf area which was only approximately half as great as the leaf area exposed to evaporation by the corn plants.

Thus, the excellent root system, coupled with a relatively small transpiring area and a low water requirement, goes far towards explaining the high degree of resistance of sorghums to drought. Their ability to remain in an almost quiescent state during drought is another important characteristic. The sorghums remain fresh and green during periods of dry weather which would be extremely harmful to corn. In fact, the plants may even cease growth for a considerable time, but when revived by a rain, a vigorous growth rate is resumed. The slow growth of the above ground parts until an extensive root system has been established is also an important characteristic in resisting drought. Drought resistance has done much to make sorghum the leading crop in the drier parts of the south and west portions of the grassland region.

Sorghums are also adapted to a wide range of soils and will thrive under conditions where other crops, like corn, do poorly. This is due in a large measure to their exceedingly well-developed root systems. Even on land that has become too poor and thin to raise corn and small grains, two or three good crops, of sorghum may be grown, often without the addition of fertilizer. This may be due largely to a more thorough occupancy of the soil, especially of the deeper soil, by the sorghum plants than by other crops.

Tillage and Crop Rotations

As for corn, methods of cultivation should be practiced with continual reference to the degree of development of the root system. Since sorghum is usually grown in regions where available moisture is nearly always limited, its conservation by a surface mulch and the removal of weeds is imperative. This may be accomplished by frequent and thorough cultivation. But such cultivation should not be too near the young plants. Later cultivations must be shallow. Practically all tests show that deep cultivation does more harm than good, as measured in yield, after the roots have attained their normal spread.

As in the case of corn, listed sorghum stands dry weather better and gives better yields than a surface-planted crop, when moisture is deficient during the latter part of the growing season. This is probably due in part, to the more deeply placed root system which is further covered by each tillage operation and is, therefore, not so subject to drought as are the roots of a surface-planted crop. It is common knowledge in sorghum-growing regions that crops following sorghum are often less thrifty and less productive than the same crops following corn or wheat.

Sorghums are believed to be "hard on the land." Sometimes, this injurious effect, is observed for more than one season. For example, in Kansas, winter wheat grown after a crop of kafir has yielded, on an average, 3 bushels less per acre per year during a period of 6 years than wheat grown after corn. This depressing effect upon yield has been thought to be due to the fact that the extensive root system of the heavy yielding sorghum depletes the soil so thoroughly of nutrients and water that it leaves it in a very poor condition as regards texture and as an abode for nitrifying bacteria. Recent investigations indicate that the depression in wheat yield is due to the harmful effects of decomposition products arising from the decay of stubble and roots after the crop is harvested.

5

Meadow and Pasture Grasses

The use of grasses for forage is scarcely less important than their use for the production of human food. Experiments in re-seeding and improving depleted range lands have shown that a knowledge of the root habits of the seedlings, as well as those of the mature plants, is of very great importance. Large amounts of forage cannot be grown nor flower stalks and seeds produced except by plants with well-developed root systems which not only absorb sufficient water and nutrients but accumulate an abundant food supply which is again used at the end of the season of dormancy.

Brome Grass

Brome grass (*Bromus inermis*) is a perennial with creeping rootstocks or rhizomes and fine fibrous roots. It is an excellent pasture grass in the semiarid regions from Kansas to Canada and westward to Washington, in which regions many eastern grasses will not thrive.

Root Develcpment in the First Year

Plants grown at Lincoln were 14 inches tall when only 11 weeks old. The roots were abundant to a depth of 2.3 feet;

a few penetrated beyond 3 feet in both lowland and upland silt loam soils. When 4 months old (Aug. 20), and the flowering stalks were beginning to appear, the grass had spread by rhizomes and produced a dense sod. The roots had formed a close network, quite filling the soil to a depth of about 2 feet. They were dark brown in color and densely clothed with root hairs, and had a great network of fine, wellbranched rootlets. The latter were several inches in length, and the tertiary branches were 1 to 2 inches long and very numerous.

Some light-coloured roots were intermixed with the brown ones. These were younger, shorter, and often less branched but took on the darker color and other typical root characters when more mature. Below 2 feet, roots were much less abundant. In fact, at a depth of 1.5 feet, they began to thin out considerably but were fairly numerous to the working level at 2.7 feet. Some extended 1 to 2 feet deeper. In the upland, the stand of plants was thinner and the roots were not quite so long.

Root Habits of Older Plants

Three-year-old plants at Manhattan, Kan., were excavated from a fine, compact loam underlaid at a depth of about 12 inches with a somewhat gravelly and sandy subsoil. This gave way at 4.2 feet to solid limestone rock. Many roots extended to the rock layer where some continued growth along its upper surface. Although the root growth was densest in the first 12 inches, the soil was filled with great masses of roots and their network of branches to the rock layer. It appears that full depth of penetration was not attained, because. at Fargo, N. D., roots of this species were found to penetrate to a depth of over 4 feet when only 1 year old and to a depth of 5.5 to 6.5 feet when 2 years old.

Thus, brome grass rapidly produces a wonderfully efficient and deeply penetrating root system, numerous rootlets occupying every cubic inch of the soil to a depth of several feet. Because of its deeply rooting habit, it is very drought resistant and well adapted for growing in light soils and in dry climates but does well also on good moist soils. Its excellent root system doubtless goes far towards making brome grass "the best pasture grass yet found for the prairie states of the Northwest and the Pacific Northwest."

Orchard Grass

Orchard grass (*Dactylis glomerata*) is an excellent pasture grass, especially when combined with other grasses. Its habit of growing in bunches or tussocks makes it less desirable for meadows. Plots of orchard grass were grown on both lowland and upland soils at Lincoln for purposes of root examination.

Root Development in the First Year

When 11 weeks old, an excellent stand of good growth, 1 foot high, characterised the grass in the lower area; that of the upland was much thinner and only 5 inches tall but of healthy appearance. Notwithstanding the great difference in above ground development, the working depth of roots was approximately the same, 2.3 feet. Maximum penetration was about 3 feet in both areas. By August 26, the tufts or bunches of grass were provided with roots which literally filled the surface soil with great masses of tan-coloured fibrous roots to a depth of nearly 2 feet, below which level they became fewer in number but were abundant to a depth of 3.2 feet.

The roots were tough and rather coarse. They were well furnished with laterals 1 to 3 inches or more in length, which were branched to the second order. The ultimate rootlets were very fine. In the deeper soil, the branching was rather largely confined to one plane in the soil crevices. Some of the delicate roots reached a maximum depth of 4.4 feet. The plants in the upland plots had developed so poorly that they were not further examined.

At Geneva, N. Y., where heavy clay loam was underlaid at a depth of 6 to 10 inches with a tenacious gravelly clay, orchard grass roots were examined during their first year of growth. Fine fibrous roots filled the soil to a depth of 1 foot, a large number reached the 21-inch level, and a few were traced to a depth of 3 feet. Some of the fine roots spread horizontally more than 21 inches from the plant.

Root Habits

Plants 2 or 3 years old at Manhattan, Kan., reached a depth practically the same as the extreme height of the grass, which was 3.5 feet. The surface 6 inches of fine, compact, sandy loam was so filled with the thick fibrous mass of roots that it was very difficult to wash out all of the soil. Below 10 inches in depth, the roots were fewer, the largest part of the root growth lying within the surface foot of soil, The deeper subsoil was quite sandy. The root system bore a close resemblance to that of wheat and oats but had an even greater fibrous growth in the surface soil. The roots were tough and woody and nearly white in color.

From these data, it may be seen that the roots of orchard grass develop rapidly, reaching depths of 2 to 3 feet by midsummer of the first season's growth. Later, as the plants continue to tiller and form the uneven sod, many new roots develop. Meanwhile, the older ones have extended well into the third and fourth foot of soil. All of the roots are well branched and furnish such an excellent absorbing system that, when pastured, this grass remains green even during a long, hot summer and late into fall.

Meadow Fescue

Meadow fescue (*Festuca elatior*) is a common European forage grass. Like the preceding grasses it has been introduced from that country into the United States, but although well adapted to the same region as timothy and bluegrass it has never been extensively grown and holds a rank of minor importance among forage grasses in America.

Root Development in the First Year

The root development of this grass was examined on lowland silt loam soil at Lincoln on June 13, when the plants were 8 inches tall but only 7 weeks old. The fine fibrous roots already had a working depth of 18 inches, and some roots extended to a maximum depth of over 2 feet. A second examination was made when the grass was nearly 4 months old and 8 to 10 inches tall. No flower stalks had appeared. The roots had a working depth of 3 feet, although some penetrated 8 to 12 inches deeper. The soil was quite well filled to a depth of 1.5 feet with great masses of brown roots, which were only slightly less abundant to 2 feet.

The roots were very fine, the largest scarcely a millimeter in diameter and many were only one-fourth as thick. They originated in great numbers from the base of the plant, 6 to 24 or even more from a single stem. Although most of the roots pursued a course somewhat obliquely to almost vertically downward, others spread laterally more or less parallel with the soil surface or obliquely outward 3 to 6 inches before turning downward. They were exceedingly well furnished with thread-like laterals ranging from a few millimeters to several inches in length, all of which were branched and profusely rebranched, ending in hair-like termini. Many of the roots did not reach the general, working level; others penetrated far beyond. The 6 to 8 inches of root ends were not so well branched and were hairlike.

Plants of meadow fescue at Geneva, N. Y., during their first season of growth, had roots which filled the soil to a depth of I foot and several were traced to the 32-inch level. A horizontal spread of 9 inches was ascertained. It seems clear that the excellent growth made by this species during the first summer is augmented, like that of most perennials, during subsequent years. Three-year-old plants at Manhattan, Kan., reached a depth of about 4 feet.

Bluegrass

Kentucky bluegrass (*Poa pratensis*) has escaped from cultivation in the prairie region, and has become an important species in the less arid parts of the grassland area. Its great economic importance as a lawn and pasture grass is well known.

Root Habits

This grass propagates by means of rhizomes which lie near the soil surface. They are 1 to 2 millimeters in diameter and well branched. The dark-coloured, fine, fibrous, minutely branched roots occur in such abundance that, with the creeping rootstocks, they form a dense, tough sod. Some of the roots have a wide lateral spread, often running nearly parallel with the soil surface at depths of only 2 to 3 inches for distances of 1 to 1.5 feet.

Usually, they run more obliquely or even vertically downward. Root depth appears to vary considerably with soil conditions. Undoubtedly, in and climates where frequent light sprinkling is practiced, the roots are very superficial, but in deep moist subsoils, they have been found at depths of 5 to 7 feet.

Root Development in the First Season

At Geneva, N. Y., in heavy clay loam, bluegrass, during its first year, made an excellent growth. The dark-coloured, very fibrous roots filled the soil to a depth of 12 inches, many reached a depth of 18 inches, and some very fine roots were traced to a depth of over 3 feet. The roots extended laterally about 12 inches from the base of the plant.

Habbits of Roots under Different Soils and Climates

In a light sandy soil in eastern Nebraska, bluegrass roots completely filled the soil to a depth of 2.5 feet. They were numerous to the 5-foot level, and a few reached a depth of 7 feet. The fine roots were abundantly supplied with hairlike laterals, usually less than an inch in length, but branched and rebranched to the third order. Even the root tips were well branched with rootlets 1 to 3 inches long.

In upland clay loam soil at Lincoln, the roots were somewhat less extensive, although bluegrass sod had been in possession of the area for 2 or 3 years. The greatest root development was in the surface 2 feet of soil, but roots were fairly abundant to a depth of 3.3 feet and the longest extended to nearly 6 feet. The wide lateral spread of surface rootlets, so highly developed in the sand, was much less in clay loam.

At the same station, in silt loam alluvial soil, the working depth was determined at 2.8 feet, although many roots penetrated deeper. A maximum penetration of 5 feet was found. A root depth of nearly 4 feet was determined for this species at Manhattan, Kan. At a depth of 2 feet, however, the roots were few in number, and comparatively few extended below 18 inches. These data are scarcely in agreement with those from Wisconsin where "blue grass is eminently a surface feeder whose roots so fully occupy the soil that there is no room for the roots of other plants to associate with them."

Root Growth of Cultivated Grasses

The preceding descriptions of four rather common meadow and pasture grasses illustrate in a striking manner the rapid growth and excellent development of the roots of cultivated grasses. In this respect, they are similar to the native species. Although the general plan of the fibrous root systems of all grasses is somewhat similar, exact study will reveal marked differences in rate of development, fineness of branching, depth of penetration, lateral spread, efficiency as absorbing organs, etc., all of which data are imperative for a thorough understanding of scientific crop production.

Soil Structure and Productivity

When the virgin prairie sod is first broken, the soil is mellow, moist, and rich and produces abundant crops. But after a few years of continuous cropping and cultivation, there occurs a great change in its physical condition. It becomes more compact and harder to till, dries more quickly than formerly, bakes more readily, and when plowed, often turns over in hard lumps and clods. After a clayey soil has been cropped for a long time, it tends to run together. It is very sticky when wet, but when dry, the adhesive characteristic almost entirely disappears.

The grass roots and humus which formerly held it together are decayed and gone. When loosened by the plow, it is often easily drifted and blown away. But when sowed to grass, marked improvement occurs, for grass is a soil builder, a soil renewer, and a soil protector. Covering the land with grass is nature's way of restoring to old, worn-out soils the productivity and good tilth of virgin ones.

The perfect tilth and freedom from clods, so characteristic of virgin soils, are always more or less completely restored wherever the land has supported a cover of grass for a number of years. The covering of sod prevents the puddling action of rain. As the roots develop, the soil particles are wedged apart in some places and crowded together in others. The small soil grains become aggregated into larger ones.

Each year, many of the old roots die and are constantly replaced by new ones. The soil is filled with pores of the old root channels; the humus from the decaying roots helps cement soil particles into aggregates and thus lightens and enriches it. In this way, the mellow texture of the virgin soil is restored, and the accumulation of organic debris, largely from the decayed masses of old roots, adds greatly to its fertility.

6

Sugar Beet

The sugar beet (*Beta vulgaris*) is a biennial. During the first year of growth, it stores a large amount of food in the fleshy taproot and crown or fleshy stem. Most of this surplus is used, if the beet is grown the second year, in the production of aerial shoots. The "beet" itself is largely the very fleshy upper portion of an extensive taproot. The upper part or crown is a very much shortened fleshy stem, upon the apex of which the leaves occur. The root proper may be distinguished from the stern or crown by the two opposite, longitudinal rows of secondary roots (Fig. 1).

Figure 1. Sugar beets about 2 months old: A, dry land (practically no water available in the second foot); B, irrigated soil.

The fine sandy loam soils are very dry unless irrigated, since the precipitation averages only 13 inches per year. The beets were grown in plots of very similar soil; one lot without irrigation, a second under light irrigation, and the third plot fully irrigated. Barnyard manure had been uniformly spread over both irrigated plots at the rate of 5 tons per acre before the seed bed was prepared.

Throughout the growing season, the aerial environment as regards temperature, humidity, wind, and evaporation was, nearly the same in the several plots. Moreover, physical and chemical composition of the soil was much the same (except for manuring), and temperatures at all corresponding depths were almost identical. Hence, the marked differences in root habit were due mainly to other soil differences, the controlling factor being water content with its attendant effects upon aeration and ease of root penetration. Plants in fully irrigated soils, of course, showed the more typical root habit.

Root Development

A widely grown variety of sugar beet known as Kleinwanzlebener was used. The seeds were drilled about 1 inch deep, in rows 18 inches apart, and in well-prepared seed beds. Later, the crop was thinned so that the beets were 12 inches apart in the row. The plots were subjected to shallow cultivation from time to time.

Early Root Development

When 2 months old (June 8), the plants had 8 to 10 leaves each and were 4 to 5 inches tall. The root development in the dry land and irrigated plots is shown in Fig. 1. All of the plants were characterised by a strong taproot.

In the dry land, practically no available water occurred in the second foot during April and very little during May, and the widely spreading, much-branched laterals compensated in part for the lack of penetration. In moist soil, the roots already averaged ¼ inch in thickness but tapered rapidly and in descending 3 to 4 inches lost half of their width. Beginning just below the soil surface and for a distance of 2 to 4 inches, laterals occurred in two rows, on opposite sides of the root. They were grouped in clusters of 2 to 4 and seldom exceeded 1.5 inches in length. This root zone was more extensive on the larger roots in the moist soil. Larger branches, originating from all sides of the root, occurred in the deeper soils.

Some arose much nearer to the end of the taproot in the less rapidly growing dry-land plants. In dry land, they were 1 to 7 inches in length, the longest ones being profusely rebranched, though in the irrigated soil, they were not only shorter but much more poorly furnished with laterals. Moreover, these primary branches were less numerous, 6 to 8 per inch, than in dry land, where 8 to 12 occurred regularly on an inch of taproot. Differences in lightly and fully irrigated soil consisted of lateral branches being somewhat more numerous and their sublaterals longer in the drier soil. Thus, the white, succulent, rather tender beet roots are considerably modified by environment.

Half-grown Root System

Excavations were again made and the roots examined a month later. Since practically no efficient rainfall occurred during this period, the dry land became almost depleted of available water. Only 2 to 4 per cent remained at a depth of 3 feet. The number of leaves varied from 15 to 18 and was about the same on plants in all plots, but those in irrigated soil were considerably larger., They were 14 inches high in the latter cage, where the tops had a spread of 18 inches, but only 5 inches high in dry land, where the spread of tops was 11 inches.

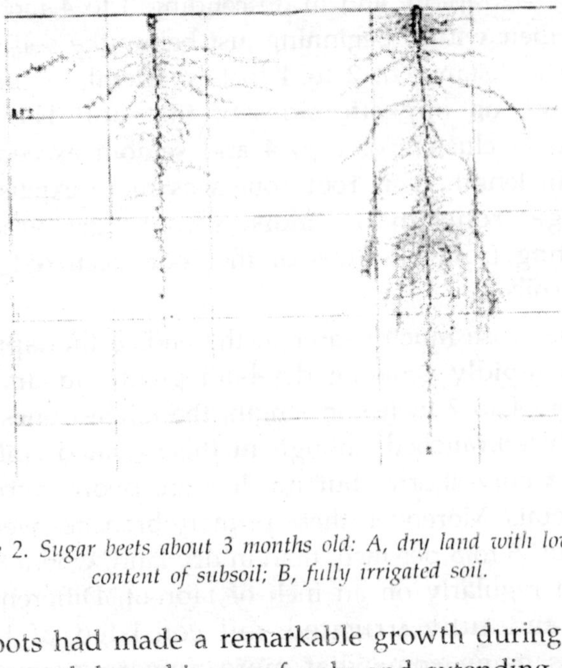

Figure 2. Sugar beets about 3 months old: A, dry land with low water content of subsoil; B, fully irrigated soil.

The roots had made a remarkable growth during the 30-day interval, branching profusely and extending into the third or fourth foot of soil (Fig. 2). The taproot had doubled in length and the absorbing area had increased many fold. Differences in development were even more pronounced than at the earlier examination. The outstanding features of the root system in the fully watered plot were the marked growth of the shallow absorbing system and the development of long, deeply penetrating branches in the subsoil. The dry-land plants had neither of these habits but were characterised by large numbers of horizontally spreading major laterals in the surface 12 to 18 inches, where soil moisture had been constantly more abundant.

Among the dry-land plants, the taproots were slightly less than 2 inches in diameter, but in the fully watered plot, they were 2.5 inches thick. All tapered so rapidly that even the larger ones were scarcely half an inch in diameter 6 inches below the surface. In the hard soil of the dry land, they zigzagged downward to depths of 30 to 46 inches, but in moist soil, they pursued a more even vertical course, reaching a depth of 50 to 54 inches.

Branches in the surface soil were very limited in dry land; only tiny rootlets less than an inch in length occurred. Their growth had apparently been stimulated by very recent showers, and they were not yet clothed with root hairs. In striking contrast, the beets which had been irrigated about 9 days earlier had developed 60 to 75 roots per linear inch in zones 4 to 5 inches long on two sides of the taproot. They spread horizontally for distances of 3 to 5 inches and were profusely branched with secondary laterals at the rate of 8 to 12 per inch.

This portion of the root system, which had failed to develop in the drier soil, added materially to the absorbing area of the rapidly developing plants. The smaller dry-land plants had many strong, widely spreading laterals in the surface foot, a response, no doubt, to the very low water content of the deeper soil. These became progressively younger and shorter downward, but the number throughout the course of the taproot was usually 4 to 7 branches per inch. The sublaterals were also long and rebranched.

In the moist soil, however, long branches occurred regularly in the first, second, and third foot, some being over 2 feet long. The direction of these roots away from the horizontal was marked. Like the taproots, they were profusely branched and rebranched with both long and short laterals.

The roots of the lightly watered beets were in many ways intermediate to those described. The course of the taproot was somewhat irregular but not to so great an extent as in the dry land. The surface roots were, in extreme cases, an inch long though, as a whole, poorly developed since water in the surface soil was scarce, but the larger branches ran more horizontally than those of the more normally developed fully watered plants.

Mature Root System

A final growth of the beets was made in mid-September. Heavy showers had somewhat replenished the soil moisture, although in the unirrigated plot, very little water was available at any time. Each plant possessed 20 to 30 leaves. Those in the irrigated plots were twice as long (about 18 inches) as those in dry land and the photosynthetic area was three to four times as great. Of course, those in the fully watered plot were largest. Effects of drought were very evident. Some of the plants in the dry land had died, and on the rest, many of the outer leaves were dead and the others frequently wilted so badly that they did not recover even at night.

The spread of the tops of these drooping plants was 19 inches, As was also that of the lightly watered plants, but in the other plot, it was 23 inches. In the fully irrigated plot, some of the older outer leaves were dead or drying, but this condition was more pronounced in the lightly watered area. Root development was correlated with that of tops. The fleshy taproots were 3 to 6 inches in diameter in the irrigated soil but only about 2.7 inches in the dry land. Moreover, the dryland plants had increased their depth from the third-foot level to about

3.5 feet (maximum 4.5 feet) since the last examination. Those in the irrigated plots, which were about 4 feet deep, now occupied the fifth and a part of the sixth foot of soil (maximum depth 6 feet, Fig. 3).

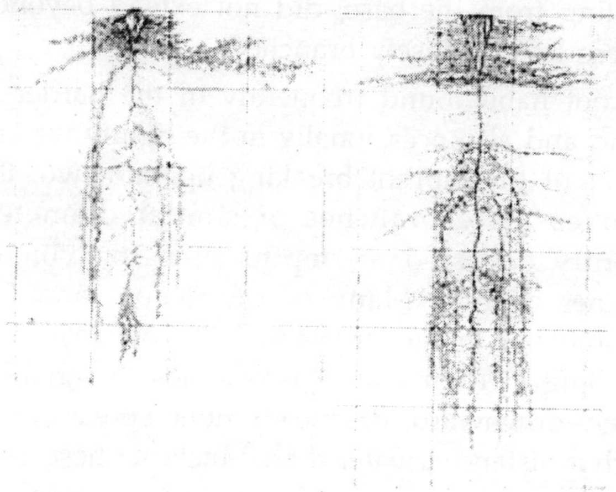

Figure 3. Sugar beets on Sept. 12: A, dry land; B, fully irrigated soil.

The root plan of the irrigated plants, with their widely spreading, horizontal surface root system and rather vertically descending and deeply penetrating major branches, was fairly well blocked out at the preceding examination on July 7. Marked changes had occurred in the root habit of the dry-land plants. Here, the two zones of surface laterals, which had then just appeared, were now well developed.

At a depth of 1 to 7 inches, in response to the increased surface moisture due to rains, these roots arose in great profusion from two sides of the taproot and ran horizontally for distances of 3 to 7 inches. Near the lower edge of these zones , they were supplemented by larger and longer roots. These spread 10 to 34 inches and were rebranched to the fifth order at the rate of 10 to 14 rootlets per inch, the whole forming a close network in the dry soil. On the larger roots of the irrigated plants, these lateral root-producing zones were longer, as were also some of the smalier roots, but the larger ones, originating from the base, did not extend beyond 2 feet and were less profusely branched.

A root habit found frequently in the harder soils of dry land and also occasionally in the lightly watered plot was that of the taproot breaking up into two, three, or sometimes more branches of similar diameter. This frequently occurred at depths of 6 to 8 inches and sometimes deeper. Many of the deeper and formerly short branches in the dry land had now reached 2 to 3 feet in length. Their course was almost invariably rather obliquely downward, practically none spreading laterally through a distance greater than 7 inches. These, with their profuse branches, occupied, more or less completely, the third and sometimes a part of the fourth foot of soil. In the irrigated plots, the roots usually spread more widely, in some instances more than 2 feet.

Seepage water occurred late in the season at a depth of about 5 feet in the fully irrigated plot, and the root penetration was not so great as in the lightly watered one, probably because of deficient aeration. Branching was similar in the two plots, except for a greater tendency for a wider lateral spread of roots in the shallower portion of the soil of the lightly irrigated plot. The sublaterals were longer, sometimes reaching 25 inches, while none were found to exceed 8 inches in the more moist soil. The surface root system of the fully watered plants contained a greater number of roots in the furrow slice, but they were hardly as long as in the drier soil of the lightly irrigated plots. The beets in the dry land, harvested late in September, yielded at the rate of only

2.5 tons per acre, those in the lightly watered plots, 21 tons, and those in the fully watered plot, 22.5 tons per acre.

Root Growth during Increased Rainfall

During the following, much more humid season the young beets in all the plots were found to have root systems practically the same and were thus in accord with the uniformity of soil moisture. Later, because of a constantly good water supply, the surface laterals spread much more widely in all plots than during the preceding year, and correlated with this was a lesser depth of penetration. As the surface soil in the unirrigated plot became drier, a pronounced tendency was found for the branches to turn downward. As a whole, the root habit more nearly resembled that in the irrigated plots than that of the last year's dry-land plants. The lightly watered soil was sufficiently wet so that little difference was found between the root habit here and in the fully watered plot. In both cases it resembled that of the fully watered plants of the preceding season.

Tillage Practice

A consideration of the very fleshy portion of the taproot explains why a deep, mellow, easily moved soil is essential for a proper development of the beet. The actual depth of cultivation varies, of course, with the nature of the soil and the previous depth of cultivation. Early tillage prevents the formation of a crust on the soil surface and subsequent difficulty in the emergence of the shoot, as well as in aeration. The time and manner of thinning is closely connected with root injury. Thinning should always be done before the seedlings are too far advanced, usually in the fourth-leaf stage.

Later thinning greatly disturbs the roots of even the more vigorous plants, left to mature. The widely spreading, deeply penetrating root systems are undoubtedly In important factor in competition and resultant reduced yields where the crop is grown too thickly. The extensive root habit explains why a deep soil is best for the growth of this crop and why a soil 2 to 3 or more feet deep and free from hardpan and standing water is absolutely essential. Since beets can use generous amounts of water to advantage and since an ample moisture supply is essential for the production of large yields, enough water should be added at each irrigation to promote good root growth. The water should moisten the soil as deeply as the roots penetrate. Cultivation following every irrigation prevents the surface soil from baking, forming a crust, and cracking with resultant loss of moisture and "pinching" of the beets. Cultivation should be shallow so as not to disturb the roots in the surface soil.

Root Growth of "Mother Beets"

Since the sugar beet is a biennial it does not form seed during the first year of growth. In temperate climates it is necessary to protect the food-stored "beets" from freezing by removing them from the soil and storing them during the winter. Beets planted in April and May showed some enlargement of the old root and especially a marked development of new side roots. On beets planted in March the side roots were numerous and large while on those planted later in the season they were few and very much finer.

The difference in root development was correlated with soil and air temperatures, the lower temperatures of early spring being more conducive to a better growth. This was shown further by the fact that "beets" planted in September when soil and air temperatures were again lower showed extensive root development. New roots are necessary to supply the demands for water made by the new shoots. They also undoubtedly absorb food materials from the soil which supplement the supply accumulated the preceding year and thus promote a good growth of tops and abundant seed production.

7

Alfalfa

Alfalfa (*Medicago sativa*), also known as Lucerne, Purple Medick and Trefoil, is a perennial flowering plant cultivated as an important forage crop. Alfalfa lives from five to twelve years, depending on variety and climate. It is a cool season perennial legume, growing to a height of 1 meter. It resembles clover with clusters of small purple flowers. It also has a deep root system sometimes stretching to 4.5 metres. This makes it very resilient, especially to droughts. It has a tetraploid genome.

Usually, it lives only 5 to 7 years, but in semiarid regions, fields over 20 years old are found. Common alfalfa has no rhizomes, although they occur on some of the hardy, yellow-flowered varieties of Medicago falcata. They also sometimes appear on some variegated types, i.e., hybrids of ordinary and yellow-flowered alfalfa. Like the wild legumes of the grassland, alfalfa is deeply rooted. The seedling usually gives rise to a single taproot which takes a vertically downward course. Considerable variation occurs in the number of side roots which arise from the taproot.

Early Root Development

A 63-day-old alfalfa plant is shown in Fig. 1. The crop

was grown on upland silt loam soil near Lincoln. The great extent of root in relation to top is very striking. When the plants were only a month old (May 1) and the tops were half an inch high, the taproots were already 5 to 6 inches long.

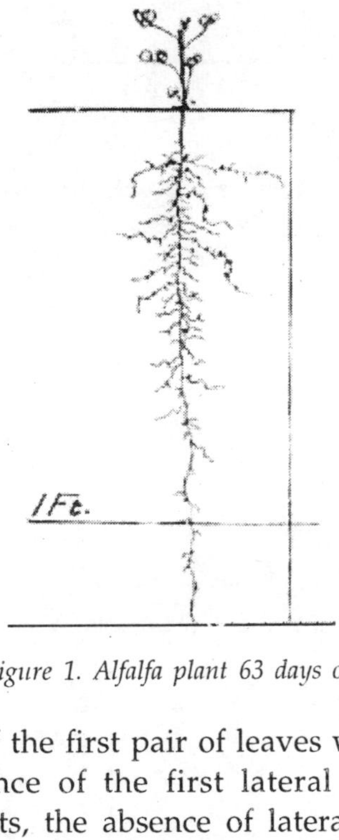

Figure 1. Alfalfa plant 63 days old.

The unfolding of the first pair of leaves was accompanied by the appearance of the first lateral roots. In the 2-months-old plants, the absence of laterals in the surface soil was as characteristic as was their abundance deeper, where 10 to 12 rootlets regularly occurred on an inch of the taproot. Some short secondary branches were present and nodules were abundant.

Later Development

By August 10, the roots had reached the stage of development shown in Fig. 2. Owing to a favourable season, the plants had made a fair growth and were 15 inches tall. Roots were fairly abundant to 5 feet, and some had penetrated to a maximum depth of 5.5 feet. The taproot was the prominent feature, laterals not being abundant below the surface 18 inches of soil. The longest branches did not exceed 6 to 10 inches in length.

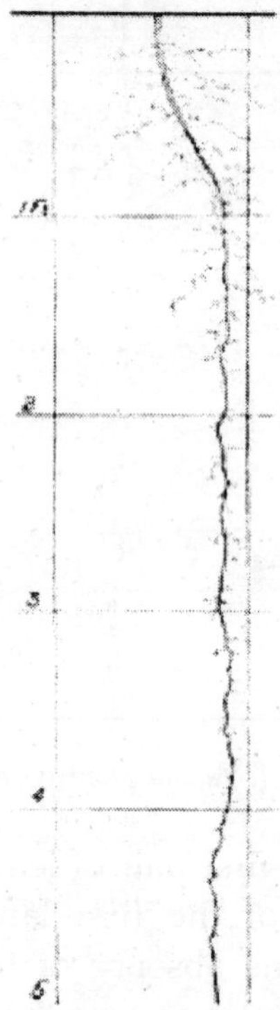

Figure 2. Alfalfa root 4.5 months old.

The general course of the roots was vertically downward. Nodules occurred to depths of over 3.5 feet. A crop of similar age on lowland soil had a better development of tops, which averaged 21 inches tall, and the roots were a little more extensive but otherwise similar. Root development in both upland and lowland during the preceding year was very similar to that just described.

Root Development of Two-year-old Plants

The lower plot was adjacent to a 2-year-old field of alfalfa. A long, deep trench was dug in this field, and the roots were thoroughly examined during the first week in June. No other plants were present, and the soil was occupied entirely by the roots of alfalfa. The taproots, near the soil surface, were 5 to 10 millimetres in diameter. Just below the crown and to a depth of 1.5 feet, the roots were well supplied with a great abundance of small laterals, usually less than a millimetre in diameter. These often ran parallel with the soil surface for a distance of 1 to 12 inches or even more. Other branches ran more obliquely downward, usually making wide angles with the taproot. They were well supplied with secondary laterals, mostly less than 3 inches in length.

All of the rootlets, but especially those in the surface 2 feet of soil, were abundantly furnished with nodules 2 to 3 millimetres long and 1 to 1.5 millimetres in diameter. In fact, these occurred to the maximum depth of root penetration, about 12 feet. The taproots tapered rather rapidly, so that at a depth of 2 feet, the diameter was seldom greater than 1 to 3 millimetres, and below 9 feet none of, the roots were more than a millimetre thick and usually much less. Many ended at depths of 7 to 10 feet, and others extended to the water level at 12 feet, where they terminated with little branching.

Not infrequently, the main root ran for distances of 2 to 5 inches in the deeper so il, giving off few or no branches. Branches more than an inch in length were rare in the deeper soil, and usually they were very much shorter. Here, the roots showed a marked tendency to branch only in the soil crevices. As a whole, the taproot predominated throughout, and typically it branched but little, many plants penetrating deeply without giving off any large laterals (Fig. 3).

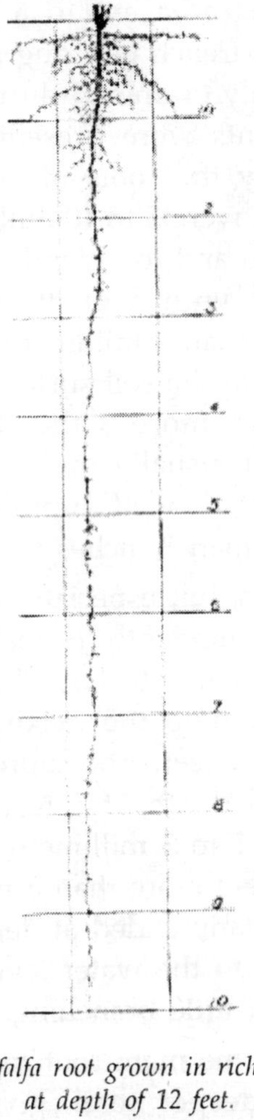

Figure 3. Two-year-old alfalfa root grown in rich lowland soil. Water table at depth of 12 feet.

The actual number of roots in a vertical layer of soil extending only 4 inches into the wall of the trench, together with the lack of large branches, is shown in Fig.

4. The presence of earthworm burrows in the deeper, stiff, clayey subsoil is significant, These, with countless small holes left at all depths, even to 12 feet, upon the death and decay of older alfalfa roots, are very important in aiding soil aeration. It would seem that the excellent development of other crops upon old fields of alfalfa, sweet clover, and red clover, the roots of which penetrate deeply, is due not only to an increased nitrate supply but also to better aeration.

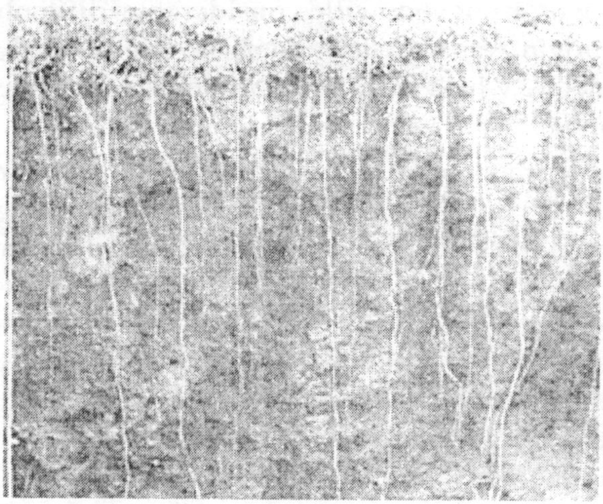

Figure 4. Upper portion of 2-year-old alfalfa roots in their natural position in the soil.

The fertilizing effect of the deeper portions of these decayed root systems is below the reach of most crops.

Effect of Environment

One part of this field of 2-year-old alfalfa extended up a rather steep hillside to the crest of the hill. Here, the plants had a good even stand, which was almost as dense as that on the low ground, and the tops were equally well developed. It was discovered that subsoil conditions were far from typical for the region. The surface 1.2 feet of dark clay loam was rather rich in humus, but below this was a subsoil of stiff yellow to slate-coloured clay about 2 feet thick. It was somewhat intermixed with streaks of decomposed Dakota sandstone which modified its tenacity. Below 3.2 feet, the clay became very hard and much jointed, roots being largely confined to these joints. The clay was intermixed with pockets and streaks of chalk. The soil was so hard, especially the deeper soil, that it was removed with considerable difficulty. Its glacial origin was shown by pebbles, often 2 to 3 inches in diameter, which occurred throughout, although not abundantly.

The taproots were as large in diameter as those from the lowland, but a marked difference in branching-habit was apparent. Branches were not only very much more numerous but much larger (Fig. 5). In fact, the first 8 inches of taproot gave rise to two or three times as many branches. Some of the larger ones ran obliquely for 1.5 feet from the base of the plant in the surface 18 inches, and then, turning downward, reached depths of 5 to 6 feet. Frequently, they divided into a large, profusely branched, absorbing network, especially prominent in the first 2 feet of soil. Very rarely did any taproots extend beyond 7.5 feet in depth, and all but the deepest were well branched, especially throughout the last 18 inches of their course.

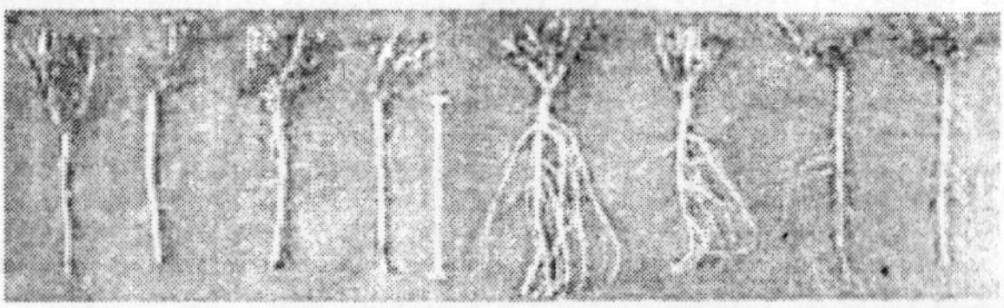

Figure 5. Portions of the root systems of alfalfa from lowland (left) and upland (right). Scale 1 foot.

These differences in root habit, namely, a deep taproot with no major branches and relatively few smaller ones in lowland and a shallower taproot with numerous large, widely spreading branches in the upland, must be attributed to soil conditions with their resultant effect upon water content, soil solutes, and perhaps aeration. Which of these or what combination of these is the controlling one can be definitely answered only by a series of experiments carried on under conditions where the effect of each factor can be evaluated. Experiments with several varieties and strains of alfalfa grown at Ithaca, New York, show that in compact soil all varieties and strains develop branch roots, but in open soil the taproots predominate.

Year-old plants grown in fairly moist loess soil in north central Kansas, from the same lot of seed used at Lincoln, had a root habit like that described for plants of a similar age at Lincoln. But those grown on the short-grass plains of eastern Colorado were very different. Although the tops were 11 inches tall and had blossomed, root penetration was limited by a hardpan layer of very dry soil which occurred at a depth of 2 feet. The root habit was very similar to that of the 2-year-old plant shown in Fig. 6. Low water content of soil and very dry air caused the crop to pass many days in a semiwilted or wilted condition, growth being resumed when showers occurred.

The taproots were only 2 to 4 millimetres thick but profusely branched with both large and small laterals. Not infrequently, some of these were equal in size to the taproot. A horizontal spread of 1 to 1.5 feet was characteristic, some roots running 2 feet in the dry soil. Small branches were so numerous that the soil was remarkably well occupied by a network of roots, a condition quite unusual in fields of young alfalfa of more humid regions. Indeed, the modification of the root habit was so great that one would scarcely recognize the roots as those of alfalfa.

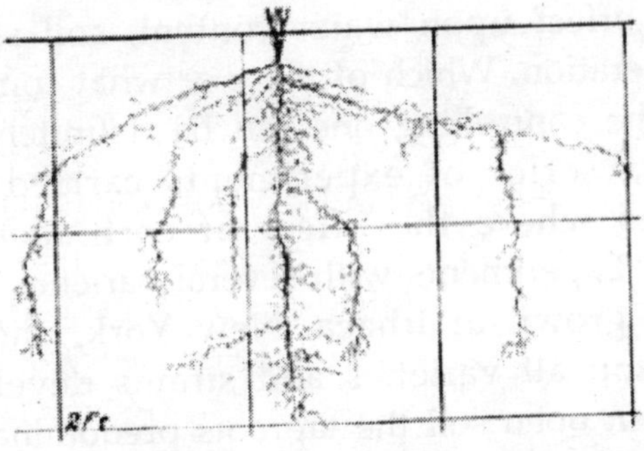

Figure 6. Alfalfa excavated in the short-grass plains on June 28 during the second year of its growth.

Chemical analyses showed that the soils were rich in all the necessary nutrients; nodules occurred on the roots at all levels; aeration could not have been a limiting factor to growth in this dry soil; and, undoubtedly, water played the dominant rôle.

Root Behaviour under Irrigation and in Dry Land

A study of crops growing with and without irrigation was made at Greeley, Colo. The plots adjoined those already described for corn. Turkestan alfalfa was drilled 1 inch deep in rows 20 inches apart on the dry land but sowed broadcast in the irrigated plot where it was worked into the soil with a hand rake. The dry-land plot was hoed at frequent intervals throughout the season to conserve the moisture, but the irrigated plot was not tilled. Sowing was done on April 11.

Early development

Owing to favourable temperatures, the crop grew well, and when 2 months old, that in the irrigated and fertilized soil was 6 inches tall. Each plant had 6 to 8 leaves. In the dry land, owing largely to a lack of moisture, the plants were only half as tall, although they had the same number of leaves. Differences in root habit were already pronounced (Fig. 7).

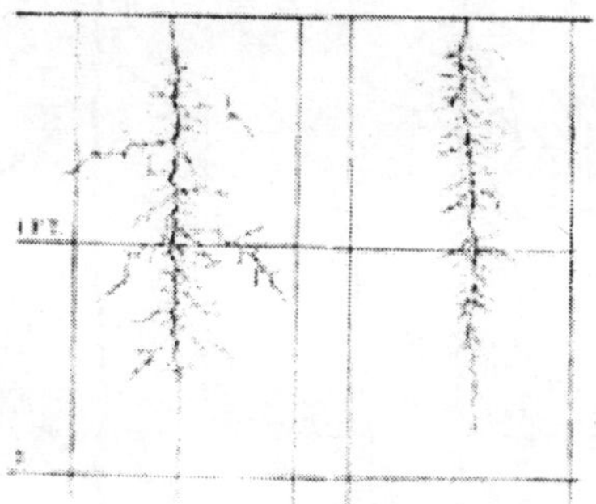

Figure 7. Roots of alfalfa plants 2 months old, those from dry land with greater depth and longer branches.

Plants in the dry land, where water was very scarce in the first foot, had penetrated deeper. They had slightly fewer but longer laterals and more secondary branches. Tubercles were also more abundant.

Midsummer growth

Root development was again examined on July 8. Practically no efficient rainfall had occurred up to this time. In the dry land, there was almost no water available in the surface foot and only 2 to 3 per cent remained in the deeper soil layers. In the irrigated field, the 3-months-old plants were 18 inches high, and many were in blossom. In the dry soil, they were only half as tall and most of them had not bloomed. A glance at Fig. 8 shows the marked differences in root habit. The prominence of the taproot, its greater depth of penetration, and the relative scarcity of large branches characterized the Plants in the moist soil. This contrasted sharply with the shallower, more profusely branched taproot found in the dry land, where several of the major branches frequently reached depths nearly or quite as great as the main root.

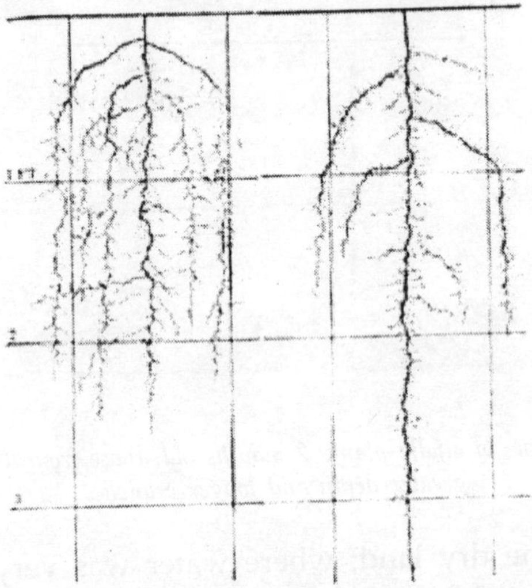

Figure 8. Roots of alfalfa plants 3 months old, the one from the dry land having the greater number of branches.

Most of the dry-land plants had from three to six large branches in the first foot; about half of the plants in the moist soil had none, and many had only one, although some, especially isolated individuals, frequently had two or more. There was a greater tendency for the branches to turn downward with less lateral spreading in dry land than in the irrigated soil. The number of small laterals varied from 6 to 10 per inch and was about the same in both cases, although they extended much closer to the root tips in the dry soil. This was due, undoubtedly, to the slower rate of elongation of the main roots. In the irrigated plot, the taproots had grown at the average rate of about half an inch a day, some reaching depths of over 3 feet.

Nodules 1 to 2 Millimetres in diameter occurred abundantly over the entire root system of the irrigated plants, but in the dry land, they were smaller, not abundant, and fairly well distributed only in the first 18 inches of soil. Thus, it seems clear that an unfavourable environment not only affects crop growth directly but also indirectly through its influence upon nodule-forming as well perhaps as on other types of nitrogen-fixing bacteria. The activities of other soil organisms are, undoubtedly, also greatly modified.

Root habit at the end of the first year

A final examination was made near the end of the growing season on Sept. 12. Rainfall since July 8 had been very light, and little water had been available at any level in the dry land. This was especially the case in the shallower soil. The second growth in the irrigated plots, the first crop having been cut on July 26, was 26 inches tall. The plants were in full bloom. In the dry land, the crop was only 8 inches tall. A root depth of 6.1 feet had been attained by the more vigorously growing crop, at which level seepage water occurred.

Plants in dry land reached a maximum depth of 5.5 feet. The roots were much more kinked and curved, probably owing to the greater difficulty in penetrating the hard, dry soil (Fig. 9). Working depths were 4.6 and 3.2 feet, respectively. As before, the dry-land plants were characterized by a greater number of strong branches in the surface foot and by a marked tendency to spread but little before turning downward and penetrating deeply. The spread of branches in the better watered soil, where they frequently originated at a depth greater than 1 foot, was much more pronounced, sometimes reaching 2 feet. No nodules were found on dry-land plants, but they were abundant to 4 feet in the irrigated soil.

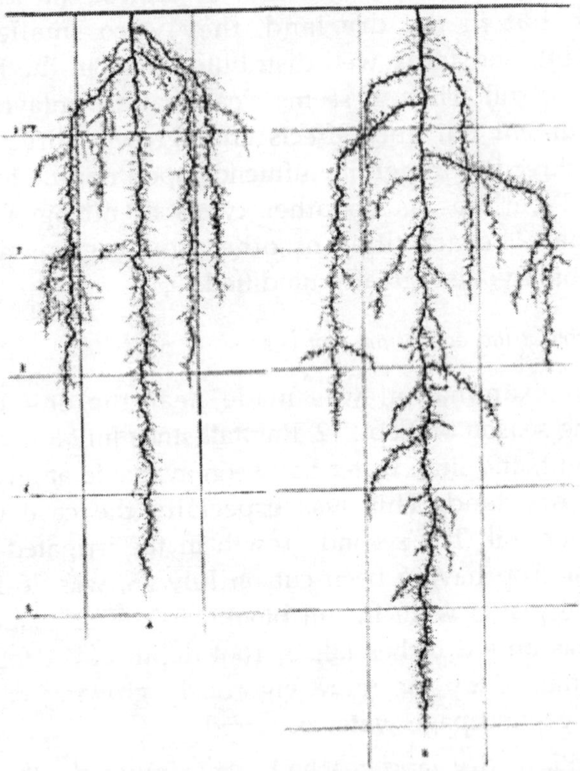

Figure 9. Alfalfa roots near the end of the first season's growth: A, dry land; B, irrigated soil.

Root extent in the second year

By July 10 of the second year, which was relatively very wet, the dry-land alfalfa had extended its depth from 5.5

feet, where dry soil had occurred, to 9 feet. Below 6 feet, the soil was sandy and gravelly. In the watered plot, growth had ceased the preceding season at the 6-foot level, owing to seepage water saturating the soil, but the roots now extended to nearly 10 feet. A comparison of Figs. 9 and 10 shows clearly the differences between the root systems in the two fields.

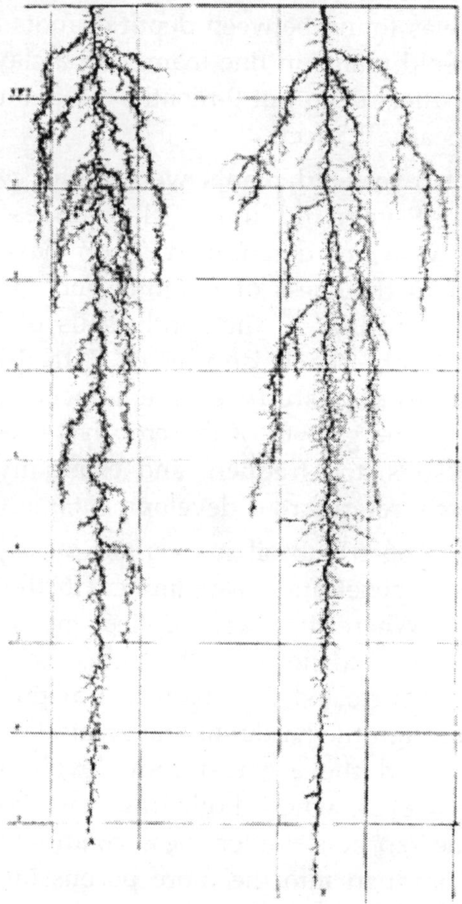

Figure 10. Root systems of alfalfa on July 2 of the second year of growth: A, dry land; B, irrigated soil.

The root habits have also been studied in several other places in Colorado in soil varying from sandy loam to heavy clay. In all cases, a marked taproot with relatively few large or widely spreading branches was characteristic. Six-year-old plants on stiff clay soil near Fort Collins were found to penetrate to a depth of about 12 feet, although 7 feet was the more usual depth of penetration found in the other Colorado soils. Little correlation was found between depth of roots and age of plants. Year-old plants in fine loam with a clayey subsoil were about 4 feet deep but 9-months-old plants had root lengths of nearly 9.5 feet.

Roots of 6-year-old plants were found which were larger than 9-year-old ones. The causes for these differences were not determined. They may have been due in part to thickness of planting and the effects of frequency of cutting. In the sandy soils of Wisconsin, where the roots reach depths of 7 to 10 feet and sometimes more, a study of the effects of different treatments on the growth of the crop has been made. In these light soils, too frequent and especially too early cutting greatly retards root development.

A porous open subsoil is overlaid by a plastic clay hardpan, root penetration was limited to the soil above the hardpan. Where lime was applied, the roots entered the hardpan but did not pass through it, though on soil which had been treated with barnyard manure, the roots extended through the hardpan. The greatest depth of root penetration and the greatest root development were attained, however, where both lime and manure were applied. The taproots, under these conditions, extended below the hardpan into the more porous lower subsoil. Increased root length of 10 to 23 inches was thus brought about. Whether the results were due to the stimulating effect of the lime and manure on the plant or to their action on the hardpan was not ascertained. 15 Undoubtedly, in soils with a very deep, moist subsoil, alfalfa roots reach great depths. However, roots seldom reach a diameter greatly exceeding 1 inch and are usually about ½ inch thick.

On upland loam soil at Manhattan, Kan., roots have been traced to depths of 8.5 feet and, in stiff clay soils, to a depth of 10 feet without finding an end. Alfalfa roots are said to extend 15 to 30 feet in depth in fairly good soil. It may be recalled that depths of 21 to 25 feet are attained by the wild rose and certain other prairie species. The need for an extensive absorbing system can be appreciated when it is recalled that the crown of alfalfa may produce from 100 to over 300 leafy stems.

Variations in Root Habit

In regard to the root habits of different varieties, studies made in South Dakota are of interest. There are outstanding differences between the root systems of southern-grown common and yellow-flowered alfalfas in the prominence of the taproots, the development of branch roots, the number and development of rhizomes, and in the number and place of most profuse production of fibrous roots. Between many plants of common alfalfa, especially of the less upright forms, and many plants of the Turkestan and Grimm alfalfas, however differences are not great, and it is often impossible to determine by their root systems the groups to which these plants belong. In brief, the root systems of the least hardy forms of purple-flowered alfalfa may be distinguished from the most hardy hybrid and yellow-flowered alfalfas with accuracy, but the intermediate forms are not sufficiently distinct to be distinguishable from one another or invariably from some forms of the non-hardy or yellow-flowered alfalfas.

Root Habit and Crop Production

An examination of seedling alfalfa roots helps to make clear why a well-prepared seed bed is essential to a good stand, the delicate roots growing best when in close contact with well-pulverized soil. It also explains why a crop sowed early in the fall does not winterkill as badly as the less developed plants of a late-sowed one. Although alfalfa is not exacting as regards soil texture and will grow well even in heavy clay soils, a knowledge of its deeply rooting habit explains why it thrives best in deep, permeable soils, as loam, silt loam, or sandy loam.

The roots are very sensitive to poor aeration resulting from standing water, complete submergence for only a day frequently being fatal, and the crop grows poorly or not at all if water stands within 2 to 3 feet of the surface. Hence, the soil must be well drained; in fact, the deeper root system demands more complete drainage of the soil than do many other field crops. If the soil lacks depth due to such factors as a high water table, hardpan, rock ledge, etc., the crop is unable fully to utilize its deeply rooting habit.

Although adapting its root habit to a considerable extent, growth will usually be less vigorous and the plant less able to withstand the competition of weeds rooting in the surface soil. Its deeply rooting habit is a character which well adapts it to semiarid regions, where, because of moderate precipitation, the soils are well drained, rich in unleached lime and other nutrients, but often fairly moist to considerable depths. Its ability to withstand drought and to continue growth, often when other plants are wilting or dead, is due, in a large measure, to its deep root system drawing upon water in the moist layers of the deeper soil.

Because of its large root growth, alfalfa exerts a very beneficial effect upon the soil, even compact soils becoming porous and friable. Aside from the fertilizing effect of the decayed roots and the increased nitrogen supply resulting from the activities of the bacteria in the nodules, some of the nutrients obtained from the deeper subsoil are left near the soil surface upon the decay of the roots and stubble and are thus made available for other plants. The roots and crowns decay quickly, leaving the soil mellow and highly productive. It should be noted, however, that alfalfa requires much more of certain other nutrients, notably phosphorus, potassium, and calcium, than equivalent yields of many other crops, and the soil is consequently more depleted of these elements.

It has been found in Colorado and in Canada that alfalfa plants having a branched root system are better able to withstand winter soil heaving than those having only a single taproot. Plants which develop rooting underground stems are able to maintain themselves after the death of the main root. When alfalfa has the habit of spreading by root proliferations, the plant is better able to recuperate from injury and to withstand cold. It has also been ascertained that such plants are generally more drought resistant. Recent physiological investigations have shown an important interrelationship between the size and composition of the roots of alfalfa and the productivity of tops.

New top growths, especially in their early stages, are initiated largely at the expense of previously deposited organic root reserves. The quality of these storage materials and their relative availability to the early growth requirements of a young shoot influence its subsequent growth and ultimate production. The translocation of the storage products from the root to the young stems results in root-reserve depletion. This deficiency cannot be restored by supplying mineral nutrients to the plant; but, since the reserves appear to be largely organic, they must be elaborated by the plant itself and translocated to the root. Before the partial depletion of the root reserves is fully replenished, the stems must reach a certain degree of maturity. In alfalfa, this is apparently the seed-pod stage.

Frequent cutting in an early bud stage checks root growth. In one experiment involving the measurement of 355 plants, the root diameter near the crown was reduced 29 per cent by two cuttings in the early bud stage as against one cutting in full bloom. The growth of new stems is accompanied by an appreciable decrease in the roots of both available carbohydrate and nitrogen reserves. With favourable soil and climatic conditions, the growth of new shoots and stems of alfalfa is largely dependent on the organic food reserves remaining in the root and crown when the last cutting is made.

Since the early growth of alfalfa stems tends to exhaust the root reserves, the primary cause of root exhaustion is a removal of the leaves and stems before they have had opportunity to replenish completely these losses. Cutting the crop too often in an immature stage ultimately robs the root and crown of its stored products, and the plant is left in a weakened and exhausted state. Each successive premature cutting results in increased detrimental effects until finally the plant dies.

Prematurely cut alfalfa often exhibits decided pathological symptoms of yellowing and stunted growth. Accompanying decreased vigour and stand is a corresponding increase in weeds and bluegrass infestation. Depleted root reserves are associated with a high water content of the root and a low concentration of dissolved organic constituents, a condition which may be a contributing factor in winterkilling.

8

Root Habits of Clovers

Clover (*Trifolium*) is a genus of about 300 species of plants in the pea family Fabaceae. They are found chiefly in northern temperate regions, but also, like many other north temperate genera, on mountains in the tropics. The plants are small annual or perennial herbs with trifoliate (rarely 5- or 7-foliate) leaves, with stipules adnate to the leaf-stalk, and heads or dense spikes of small red, purple, white, or rarely yellow flowers; the small, few-seeded pods are enclosed in the calyx.

Clover inflorescence

Eighteen species are native in Britain, and several are extensively cultivated as fodder-plants. The most widely cultivated clovers are White clover *Trifolium repens* and Red clover *Trifolium pratense.*

Habits of Various Clovers

Clovers are among the most widely known and most important of cultivated legumes. They are a basic, crop in the agriculture of the greater part of the northeastern States and as far west as the Great Plains. The great value of clovers as pasture plants and as hay crops is familiar to nearly every one. Since they are able, with the aid of bacteria in their root tubercles, to make use of the free nitrogen of the air and to store it in their tissues, their use in crop rotations is very important in maintaining soil productivity.

Common Red or Purple Clover

Red clover (*Trifolium pratense*) is a biennial or perennial plant. It is one of the most important and most widely known of all cultivated legumes. The life period is a varietal character, the average, perhaps, being about 2 years. The top develops from a pronounced taproot which penetrates very deeply and possesses an extensive system of laterals.

Early development

The root system of a plant 2.5 months old, grown in a field near Lincoln, Nebr., in rich, moist, silt loam soil is shown in Fig. 1. The top, at this time, was scarcely 12 inches tall. The strong taproot, the widely spreading and muchbranched laterals, and the abundance of tubercles are all characteristic.

By the middle of August of the first growing season, the plants had blossomed profusely, and some were ripening seed. They were about 13 inches tall and had from 3 to 13 stems from each root. Several taproots, 5 to 7 millimetres in diameter near the soil surface, were traced to depths of 4.5 feet. These tapered so rapidly that at a depth of 9 inches they were only a millimetre thick. Frequently, as many as 10 major branches arose from a single root in the surface 8 inches of soil. They ran horizontally or slightly obliquely 4 to 12 inches before turning downward, where, tapering rapidly, some of the longest reached depths of 3 to 4 feet.

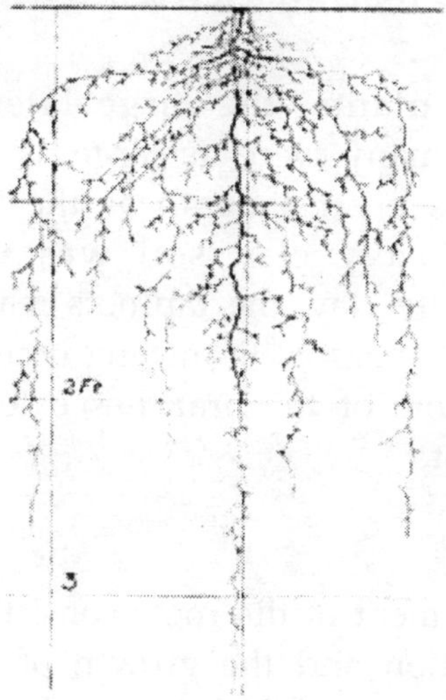

Figure 1. Root system of red clover about 2.5 months old.

In addition to these laterals, the first 6 to 12 inches of taproot were clothed with very numerous smaller branches 1 to 7 inches long. Below the first foot, no large branches occurred, although, infrequently, the taproot divided into two more or less equal parts. Short branches, only a few millimetres apart, arose throughout the course of the root to a depth of 3 to 4 feet. The joints of the deeper soil, especially, were filled with networks of rootlets. The taproot was prominent throughout and, in general, pursued a vertically downward course.

On drier upland soil, the plants were smaller and slower in blossoming. There were, also, fewer stalks per plant. The largest roots were only 5 to 6 millimetres in diameter. None penetrated deeper than 4 feet. The root habit was very similar to that described, but the larger laterals were often fewer. Here, in a rich drift soil of clay, sand, and loam, plants at the ages of 1, 2, and 5 months, respectively, had roots which reached depths of 7, 22, and 68 inches.

On all the plants, the large laterals were very abundant, and many were as large as the taproot and penetrated to as great a depth as the taproot. In New York, where heavy clay soil was underlaid with tenacious, gravelly clay, the taproots reached a depth of 34 inches the first season. Branching occurred throughout their length, some of the branches extending 12 inches from the taproot.

Mature root system

Further development of the roots consists largely in their deeper penetration and the growth of some additional laterals. A mature root system is shown in Fig. 2. A comparison with the young plant in Fig. 1 readily reveals the fact that the territory later fully occupied by the roots was already blocked out early in their growth. The fully grown plants were excavated in deep, fertile silt loam which was underlaid to a depth of at least 10 feet with a fairly compact, moist, loess subsoil. Strong taproots, nearly half an inch thick, were common but tapered so rapidly that at a depth of 1 foot, they were seldom more than half as large.

The roots penetrated nearly straight downward to depths of 8 to 10 feet. A great mass of fine rootlets arose from the crown and first few inches of the taproot and, running laterally for distances of 6 to 8 inches, quite filled the surface soil.

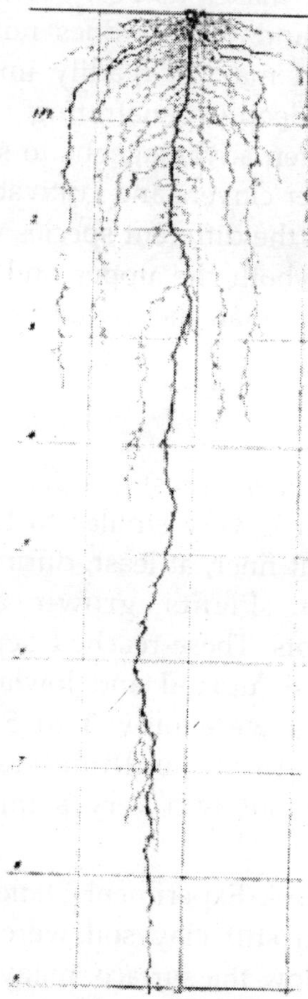

Figure 2. Mature root system of red clover.

As illustrated in Fig. 2, a few larger branches, 2 to 3 millimetres in diameter, arose in the surface foot and extended out horizontally or obliquely for distances of 0.5

to 1.5 feet before turning downward. These usually extended to the third- or fourth-foot level. Below the surface 1.5 feet , lateral roots were more sparse. In the deeper soils, the branches were very fine and ran for several inches without branching.

Why red clover makes its best growth on rich, deep, well-drained soil and why it does not thrive on low, poorly drained soil may be readily understood from a knowledge of its deeply penetrating root system. One reason why it is often advantageous to seed red clover in a mixture with other clovers and cultivated grasses is that the root systems of the different species vary widely. As a result, the soils of both the upper and lower layers are more fully occupied than they would be by a stand of a single crop.

White Clover

White clover (*Trifolium repens*) is a perennial legume with a root habit which is very similar to that of red clover, although somewhat finer, at least, during the first year of its development. Plants grown at Lincoln had pronounced taproots. These reached depths of 2.5 feet by midsummer in both upland and lowland silt loam soil, although the tops were only 3 to 5 inches tall. The taproots were 2 to 2.5 millimetres thick, and the branching throughout was very similar to that of red clover.

At the New York Experiment Station, the taproots of plants growing in stiff clay soil were found to branch about 2 inches below the surface, many roots extending 9 inches from the taproot. While the majority of the roots were within the surface 15 inches of soil, a few reached depths of 2 feet during the first year of growth. Mature plants possess long, deeply penetrating taproots from which originate many profusely rebranched laterals. Root tubercles occur throughout.

White Sweet Clover

White sweet clover (*Melilotus alba*) is a biennial crop and roadside plant. Like many other legumes, it is characterised by a strong taproot. It makes a fair growth even in soils so depleted of nitrogen and humus that they will not produce other crops profitably. It even thrives on newly exposed heavy clay soils or upon steep embankments where little else will grow. But it makes its best development in fertile, moist soils where it may reach a height of 1 foot in 7 weeks.

Early development

Seven-weeks-old plants were found to possess strong taproots which ran vertically downward to depths of 2 to 2.6 feet, although the roots were only 1 to 3 millimetres in diameter.

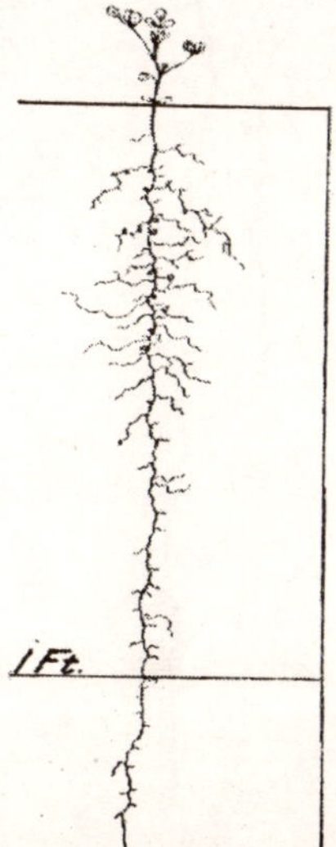

Figure 3. Sweet clover plant 63 days old, grown in upland soil.

They were exceedingly well branched throughout their course, even to their tips. The branches were small and mostly 2 inches or less in length, although some had reached lengths of 5 to 6 inches. Tubercles were abundant (cf. Fig, 3).

Four-months-old plants

When 16 weeks old, the stems were about half a centimetre in diameter and 21 inches tall. A typical root system is shown in Fig. 4. The taproots were of the same diameter as the stems. They tapered rapidly, some being a little more than 5 feet deep. Rarely more than two large branches occurred on a plant. They originated at various depths to 4 feet. In thin stands, the roots are often much more branched.

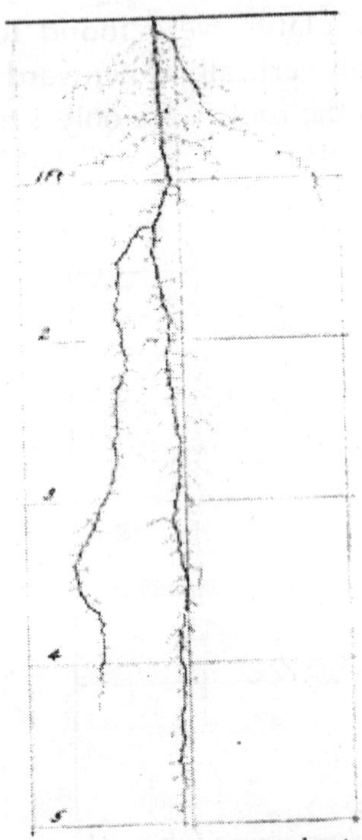

Figure 4. Sweet clover root system about 4 months old.

Many small laterals clothed the taproot and major branches throughout their course. These varied from a few millimetres to several inches in length. The younger, glistening white roots of the deeper soils branched not at all or only poorly, except where they occurred in the crevices of the jointed soil. Here, perhaps because of better aeration, they branched profusely. Nodules were of frequent occurrence to depths of 3 to 4 feet. In upland soil, plants of the same age but with smaller tops had roots which extended even deeper (6.5 feet).

A similar rapid development of the roots of this species was found at Geneva, N. Y. Here in stiff clay soil, the taproot had reached a depth of 33 inches by midsummer (July 27). Branches extended laterally 18 inches. The surface 1.5 feet of soil was quite filled with these and their sublaterals.

Twelve-months-old plants

The root system of a 12-months-old plaint is shown in Fig. 5. This grew in rather low level land adjoining a sandy ridge near Central City, Nebr. The old bluegrass sod had been covered to a depth of 1.2 feet by windblown sand from the adjacent hills, and the sweet clover had been planted to prevent further local soil shifting. Below the drifted sand, there was a layer of 1 foot of black sandy-loam, underlaid by 2.8 feet of fairly pure sand. Seepage water occurred above a layer of clay at the 5-foot level. This undoubtedly hindered further root penetration, for mature plants, in well-drained soil, may reach depths of 8 feet or more.

The numerous large branches, which, on many other plants, originate also nearer the soil surface, were very characteristic. The first 1.5 feet of the taproot had numerous small laterals only a millimetre or two in diameter. These ran horizontally to distances of 0.5 to 1.5 feet, or even more, and filled the surface sand with a delicate network of absorbing roots. Few of the main laterals spread to a greater distance than 2 feet from the downward course of the taproot, but most of them reached depths of about 5 feet. All were abundantly supplied with extremely well-branched sublaterals so that the soil was well filled with roots to this depth. Thus, sweet clover is not only quite deeply rooted but is also fitted to absorb at all levels in the soil.

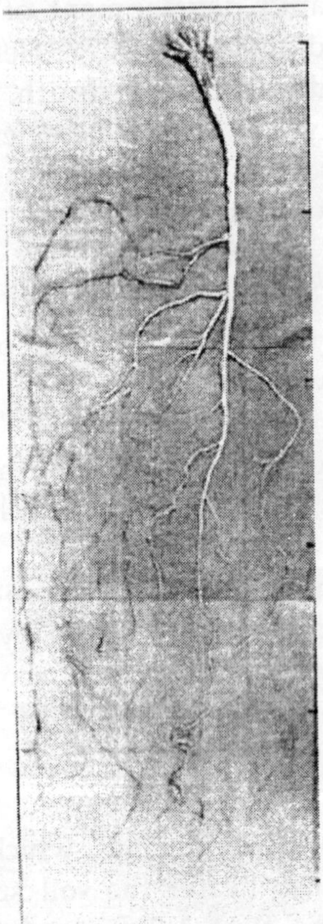

Figure 5. One-year-old sweet clover root system. Scale in feet.

The root system of the annual white variety, Hubam clover, may exceed 6 feet in length. Comparative field tests in Ohio have shown that the biennial plants have a very much larger and deeper root system than Hubam. The weight of the roots to a depth of a foot was seven times greater, and the percentage of nitrogen in the roots four times as great as in the annual variety.

The effects of roots of leguminous crops upon their death and decay in loosening and aerating the deeper soil are very great. In old clover and alfalfa fields, the soil is quite filled with root channels which greatly modify its structure and promote aeration. The amount of manure they add to the soil is likewise significant. This has been calculated from careful measurements made over small areas. In the case of a good stand of 2-year-old red clover, 6,580 pounds of vegetable matter—over 3 tons per acre—were left as roots in the soil. Chemical analyses showed that this included 180 pounds of nitrogen, 71 pounds of phosphoric acid, and 77 pounds of potash. Of course, a long series of decomposition processes is necessary before the materials in the decaying roots again become available for plant use.

9

Potato

The potato (*Solanum tuberosum*) is a perennial plant of the Solanaceae, or nightshade, family, commonly grown for its starchy tuber. Potatoes are the world's most widely grown tuber crop, and the fourth largest crop in terms of fresh produce (after rice, wheat, and maize), but this ranking is inflated due to the high water content of fresh potatoes relative to that of other crops.

The potato originated in the Andes, likely somewhere in present-day Peru or Bolivia. Potatoes form an important part of Andean culture, and the farmers grow many different varieties possessing a remarkable diversity of colors and shapes.

The root habits of the potato (*Solanum tuberosum*) have been thoroughly studied in many places. At Peru, potatoes were grown in mellow loess soil in rows 3 feet apart. The plants were spaced 2 feet apart in the rows. The field was kept clean by shallow hoeing so that the potato roots lying near the surface were not disturbed. Planting was done in April first week, and the roots were examined in May last.

Stages of Root Development

Early Development

When the tops were nearly a foot high, the root system was almost entirely in the surface 6 to 8 inches of soil. As many as 55 roots originated from the base of a single plant and ran practically parallel with the surface of the soil (Fig. 1). They varied from a few inches to 2.2 feet in length. None penetrated deeper than 1.5 feet. The soil about the plants was so thoroughly filled with these roots that it was found impossible to represent all of them in one plane. Consequently, the drawing shows but one-half of the entire root system. The close proximity of the roots to the soil surface, some being entirely confined to the first 2 inches, should be taken into account in tillage practice. Deep cultivation would have been very destructive to the roots and would undoubtedly have upset the nicely adjusted balance between root and shoot, resulting in a materially decreased yield.

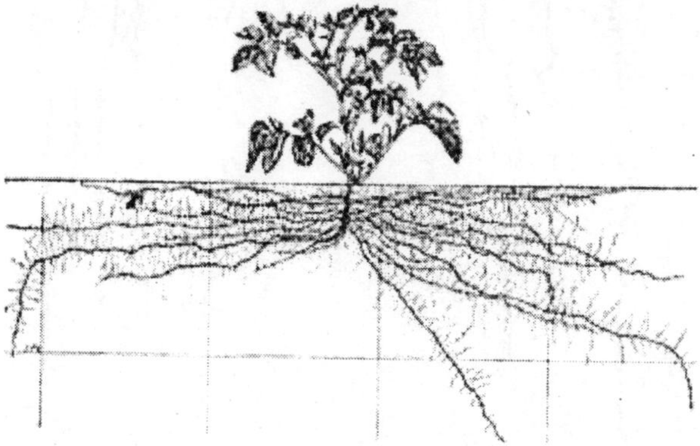

Figure 1. One-half of the root system of an Early Ohio potato plant, 56 days old.

It should be noted that a number of the longer roots showed a distinct tendency to turn rather abruptly downward. This is a marked character in the normal growth habit. The main roots were densely covered with thread-like branches varying in length from a fraction of an inch to 4 inches. These were so numerous that the surface soil was thoroughly occupied by them. Several young tubers had begun to appear.

Mature Root System

When growth was complete and about one-third of the leaves dead, the plants were again examined. This was on July 8, 94 days after planting. One-half of the mature root system is shown in Fig. 2.

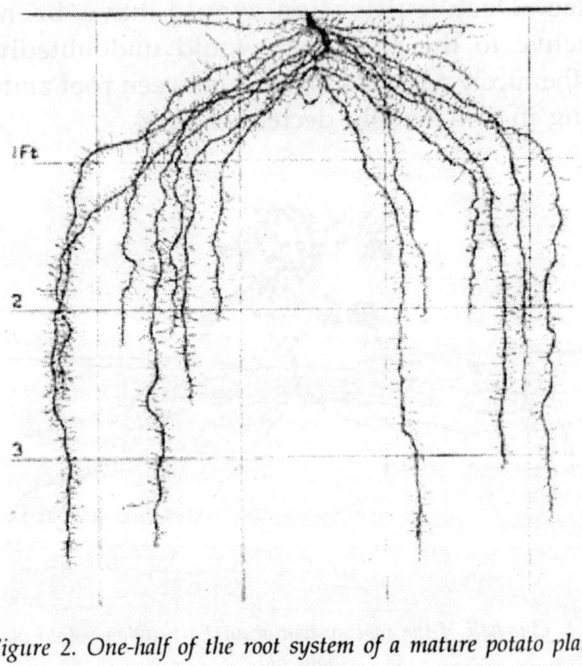

Figure 2. One-half of the root system of a mature potato plant.

The form of the root system was almost identical with that found 5 weeks earlier. Practically the only difference was in its extent. With few exceptions, the roots ran outward and downward in the surface foot and often in the surface 8 inches for distances varying from 6 inches to over 2 feet. They then curved more or less abruptly downward and continued their irregular course to a depth of 2 to 4.7 feet.

Some were still distinctly shallow, running their entire course in the surface 2 to 3 inches of soil. The paucity of roots penetrating vertically downward beneath the plant is in striking contrast to the habit of corn and the smaller cereals. All the roots, whether shallow or deep, were freely branched throughout their course, even to their tips. These fine, white, frequently rebranched laterals were usually 0.5 inch to 4 inches long, although some near the surface reached 15 inches.

Thus, in its early growth, the potato has a distinctly superficial, widely spreading root system. Later the roots turn downward and rather thoroughly occupy the second, third, and part of the fourth foot of soil. The soil volume directly under the plant is often less completely occupied. The individual plants were more variable in respect to number and extent of roots than any of the monocotyledonous crops.

During a following season, the tendency of the roots to turn downward after spreading widely was not so marked, perhaps not more than 30 per cent penetrating far below the 1.5- to 2-foot level. Drier soil both above and below the second foot may have accounted, in part, for this difference. The root ends were much more branched than previously. In all other respects, the root habit was similar.

Tillage in Development

The relation of the root habit of potatoes to methods of cultivation is apparent. It is very similar to that of corn. Thorough tillage should be given during the early development of the crop so as to afford the most favorable environment for the growth of the roots and tubers. Each successive cultivation should be shallower than the preceding one and farther away from the plants. Usually, during the final cultivation, enough soil should be worked in towards the vines to give a little ridging. This protects the tubers growing near the surface from sunburn and frost. Careless cultivation or its continuance too late in the season often causes low yields. A single cultivation when the soil is well filled with roots may, in the absence of a rain soon after cultivation, reduce the yield fully one-half.

Root Habit under Different Degrees of Moisture

Like other cultivated plants, the potato gives a marked response to differences in environment. Bliss's Triumph potato, widely cultivated in Colorado as an early variety, was grown in an un-irrigated plot of fine sandy loam soil at Greeley, during 2 successive years. As indicated by the water content of the soil, the second year was one of much greater rainfall. The roots of plants grown in the same field during the 2 years are shown in Fig. 3. The hills were spaced 14 inches apart in rows 3 feet distant.

During the drier year, no water was available in the third foot and only a small amount in the surface soil. Scarcity of water promoted an early extensive root development. The exceedingly numerous, long, much-branched laterals thoroughly ramified the dry soil. Owing to hot, dry weather, coupled with little available water, which at certain levels was practically none, the plants did not increase in size between June 23 and the next examination on July 7. They wilted almost daily and regained partial turgidity only during the cool nights. Under these conditions, growth was poor.

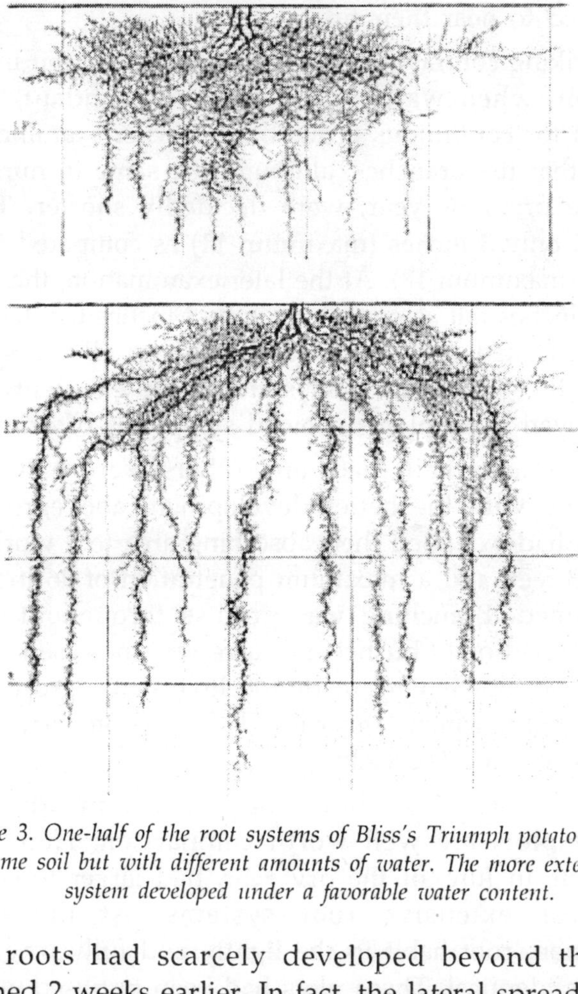

Figure 3. One-half of the root systems of Bliss's Triumph potato grown in the same soil but with different amounts of water. The more extensive root system developed under a favorable water content.

The roots had scarcely developed beyond the state reached 2 weeks earlier. In fact, the lateral spreading and branching had not changed, except that the branches now extended to the root tips, indicating the cessation of root elongation. The working depth, however, had been increased 6 inches, and the maximum penetration from 23 to 26 inches. Branching was profuse throughout. Laterals occurred at the rate of 15 to 20 per inch and ranged in length from ½ inch to 18 inches. These were thickly rebranched to near their tips.

In striking contrast was the extensive root habit in the same soil, when water, although not abundant, was sufficient for continuous growth. The earlier examination showed that the branches, although the same in number as in the previous year, were decidedly shorter. They averaged only 3 inches (maximum 11) as compared with 5 inches (maximum 18). At the later examination, the tops were 17 inches tall as compared with 8 inches for the dry year and lacked only 1 foot of occupying all the space between the 3-foot rows. In spite of the excellent root system, even these plants showed symptoms of drought.

Many leaves had died, and others were half dry. Correlating with the better development aboveground, the roots had extended their absorbing area to a working level of 3 feet, and a maximum penetration of 46 inches, was attained. Branching was profuse throughout. The previous yield of 19 bushels per acre was increased to 29 bushels, the tops having made too luxuriant a growth to resist drought, which later caused the crop to dry early and thus reduced the yield.

Root habits of similar crops grown in adjacent irrigated plots are well worth comparison. Here, the plants, late in June of the dry year, had larger tops but much less extensive root systems. At the June examination, root habit in the lightly and fully irrigated plots was identical. These plots had been fertilised at the rate of 5 tons of barnyard manure per acre before the seed bed was prepared. Root habit differed from that in the dry land in a slightly shallower penetration, fewer roots occurring in the second foot of soil, and especially in fewer and shorter branches.

At the July examination, when the plants were blooming, the fully irrigated crop was rooted mostly in the surface 16 inches of soil, although a few roots had penetrated an inch into the third foot. The lateral spread was about 2 feet on all sides of the hill as compared with 2.5 feet in the dry land. The lightly watered potatoes had a greater lateral spread than the fully irrigated ones. Moreover, the second foot of soil was occupied much more thoroughly, since more roots took a vertically or nearly vertically downward course. Branching was more profuse; the working level was 8 inches deeper and the maximum penetration 7 inches greater. The largest yield, 303 bushels per acre, was made by the lightly watered plants which had also the most extensive root systems.

Investigations at other stations indicate that potatoes frequently do not extend so deeply as pictured in Figs. 2 and 3. This may be explained, in part, by the different rooting habits among varieties although soil conditions are often equally important. At Geneva, N. Y., roots of the White Star potato reached a maximum depth of only 1.6 feet. The horizontal roots, however, were traced to a distance of 2.2 to 2.5 feet from the base of the plant.

Most of the roots were in the 14 inches of surface Soil. At Fargo, N. D., it has been found that late-maturing varieties root more freely and more deeply than early ones. For example, the Early Ohio variety, 43 days after planting, had its roots confined for the most part to 8 inches of surface soil, although a few extended to a depth of 1.5 feet, and some of the horizontal ones reached a length of 2 feet.

At maturity, the roots had penetrated to a depth of 2.5 feet. But the rural New Yorker No. 2, a late-maturing variety, reached a depth of 3 feet, the lateral roots being well interlaced between the hills, which were 3 feet apart. Investigations in Colorado have shown that, in good soil, the roots will often penetrate 2 feet in depth and extend laterally 2 to 3 feet. Depths of penetration of 4 to 5 feet have been reported for potatoes grown in loose and well-drained soils in Utah.

Experiments in which the soil layers, at various depths, were fertilised with sodium nitrate showed clearly that the root growth is markedly affected by the presence of fertilisers. Whenever the roots entered an enriched layer of soil, they branched much more freely, while growth in depth was considerably retarded. In most cases, the growing root system was 6 to 8 inches shorter in the fertilised soil, even in mature plants.

10

Sunflower

The sunflower (*Helianthus annuus*) is an annual plant in the Family Asteraceae, with a large flower head (inflorescence). The stem of the flower can grow up to 3 metres tall with the flower head reaching 30cm in diameter.

Sunflower

During the past decade, sunflowers (*Helianthus annuus*) have been frequently grown for silage in the United

States. Their wide distribution as a weed, especially in the semiarid Western States, adds further to the interest and value of a knowledge of their root habits. Plants of the Russian variety were seeded on both upland and lowland soil near Lincoln, Nebr., and their root habits studied at different stages in their development.

Stages of Root Development

Early Development

On June 2, when the plants were about 5 weeks old, they had reached a height of 15 inches in the rich silt loam of the lowland. The strong taproots penetrated quite vertically downward to a working depth of 18 inches, some extending slightly into the third foot of soil. Numerous rather widely spreading branches occurred, being especially abundant in the surface 6 inches. These spread laterally, some nearly parallel with the soil surface, to a maximum distance of about 12 inches.The older ones were well furnished with smaller branches. In the drier and less fertile upland plots, the plants were only 10 inches tall, but here, also, the extent of roots exceeded that of the tops, the deepest roots ending in the clay subsoil at the 15-inch level.

Later Development

A second examination was made when the crop was 2.5 months old. The plants, spaced only 20 inches apart, had made a vigorous growth. They were nearly 7 feet tall and had a stem diameter near the base of almost 1.5 inches. Each individual was furnished with 35 to 40 active green leaves. The flower heads were fairly well formed, and a week later, the plants would have been in bloom. The major portion of the root system is shown in Fig. 1. The roots occurred in such great numbers that it was quite impossible to represent all of them in the drawing without confusion. Hence, the front portion of the plant, with its accompanying roots, was removed before the penciled draft was made in the field. The bulk of roots occurred in the surface 1.5 feet of soil,

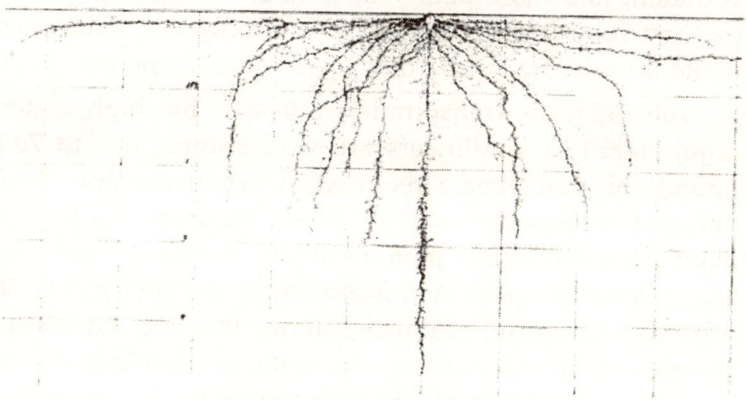

Figure 1. Root system of a 2.5-months-old sunflower.

The enlarged taproot gave off so many laterals and tapered so rapidly that, at a depth of 8 to 10 inches, it was only 4 to 5 millimeters in diameter and, in fact, no larger than some of the major branches. The taproots penetrated nearly straight downwards to a depth of about 5 feet. In the surface 6 inches of soil, 28 large laterals originated. Some of these ran obliquely, at various angles with the taproot, spreading rather widely, and reached depths of 2 to more than 3 feet.

Numerous others took a course more or less parallel with the soil surface and ran to distances of 3 to 4 feet, where they ended in the surface 6 inches of soil or, more rarely, turned downward. One large lateral was traced to a distance of 5.5 feet from the base of the plant. The main laterals gave off few or no large branches but were thickly clothed with smaller ones. The surface 10 inches of soil, especially the first 2 feet on all sides of the plants, were so densely filled with great masses of branched and rebranched roots of all sizes that they formed a complete network. Indeed, a more profusely developed absorbing system can scarcely be imagined. The roots were less abundant and more poorly furnished with laterals below the first foot. Even in the third foot, however, glistening white, poorly branched roots were quite abundant.

The vigorous transpiration rate and the high water requirement of sunflowers are well known, 600 to 700 pounds of water being required to produce a pound of dry matter aboveground. A, single vigorous sunflower plant may produce 2 pounds of dry substance and use 150 gallons of water. A comparison of the root habits of corn with those of sunflower shows that the sunflower absorbs water and nutrients from the same soil levels as the corn. In addition, the broadly expanded leaves of the sunflower shade the corn plant. Clearly, there is no place for sunflowers or other weeds in cultivated crops and especially in semiarid regions where lack of sufficient water is the chief limiting factor to crop production. The results of competition with weeds are dwarfed plants and decreased yields.

Variations in Root Habit under Competition

Roots of sunflowers, like those of other plants, are greatly modified in their development as a result of competition with other species or individuals. In one experiment, plants were grown 2, 8, and 32 inches apart, respectively, in three different plots. Competition for light occurred in all three fields, but it started earlier and was more severe in the thicker plantings. Repeated soil moisture determinations showed that the more vigorous, widely spaced plants were drawing upon the soil moisture to a degree very similar to that of the thicker plantings.

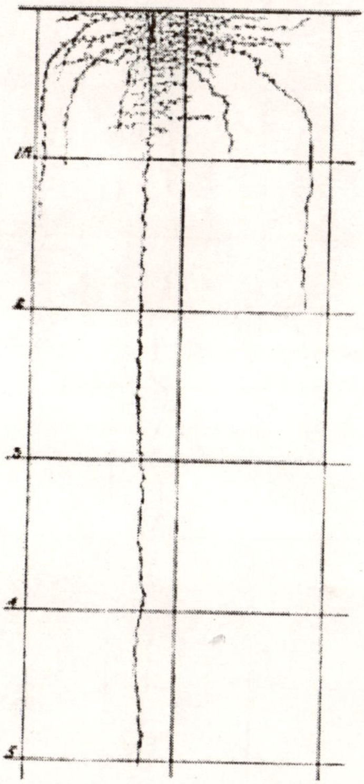

Figure 2. Root of mature sunflower where plants were spaced 2 inches apart.

A representative root from each of the thicker plantings is shown in Figs. 2 and 3, and Fig. 1, except for its lack of depth of both taproot and laterals and slightly greater lateral spread, is very representative of the 32-inch plantings. This illustrates the fact that rate of planting of cultivated plants or their competition with weeds results in pronounced modifications of both aboveground and belowground parts.

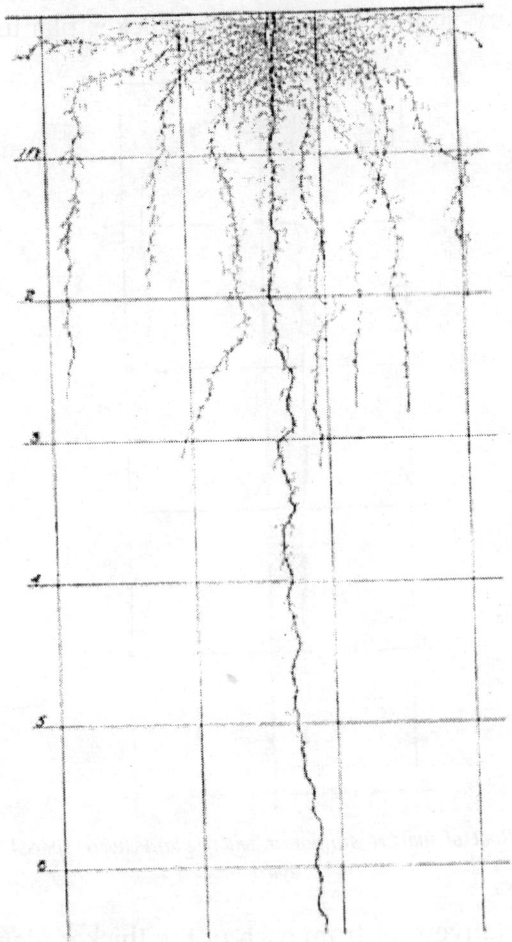

Figure 3. Root of sunflower where plants were spaced 8 inches apart.

Ample development of roots as well as tops is imperative for maximum yields.

11

Wheat

Wheat (*Triticum*) is an annual. But under the climate and cultural conditions, there are two seasonal forms, viz., spring wheat, which is a summer annual, and fall or winter wheat, a winter annual. Both have a fibrous root system which penetrates deeply into the subsoil. That of winter wheat, perhaps because of the longer season for growth, is more extensive. Upon germination of the grain, the primary root takes the lead, but very, soon, two other roots appear on opposite sides of the first. To this whorl of three, still others may be added, and together they constitute the primary root system. In some cases, there may be as many as eight roots. Early in the development of the plant, roots of the secondary root system grow from nodes above the primary one. The first whorl of roots of the secondary root system always develops within an inch or two of the soil surface. The number of roots increases somewhat in proportion to the number of tillers.

Stages of Wheat Development

Early Development

On May 1, a month after planting and when the second

leaf was half grown, a typical root system wasdrawn (Fig. 1). The number of roots varied from three to eight. Lateral roots were fairly abundant but entirely unbranched. The greatest lateral spread was 5 inches and the working depth or working level (i.e., a depth to which many roots penetrate and to which depth considerable absorption must take place), 6.5 inches. Compared with the shoot development, the plant had made an excellent growth underground.

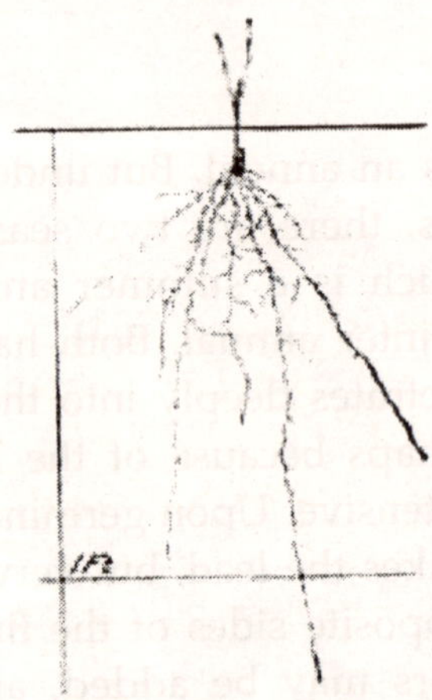

Figure 1. Marquis spring wheat 31 days old.

A plant 45 days old is shown in Fig. 2. During the 15-day interval since the first examination, two tillers and four or five new roots had developed on most of the plants. Young roots only 1 to 2 inches long were frequent. Thus, the balance between transpiration and absorption was well maintained. Lateral spread had increased 4 inches and working level about 3.5 inches. Moreover, lateral branches were much longer and secondary branches were beginning to appear. The slow rate of growth is shown by the fact that wheat planted in the same field on May 5 had a more advanced growth when only 25 days old.

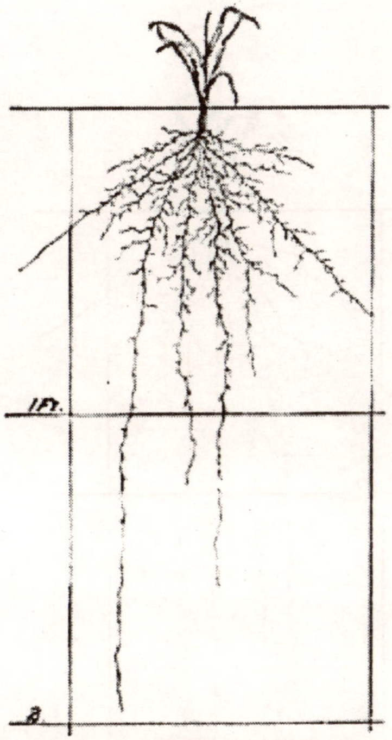

Figure 2. Wheat 45 days old.

Half-grown Plants

Fifteen days later when the plants were 2 months old, the root habit was again examined (Fig. 3). The crop was now 8 inches high and the parent stalks had four to six leaves. Tillering had increased, and about five new roots, on an average, had been added to the secondary root system. The lateral spread had increased but slightly. Many roots had penetrated deeper, however, and others had spread obliquely downward and, with the increase both in number and length of laterals, had begun to fill in the soil volume already delimited at the earlier stage.

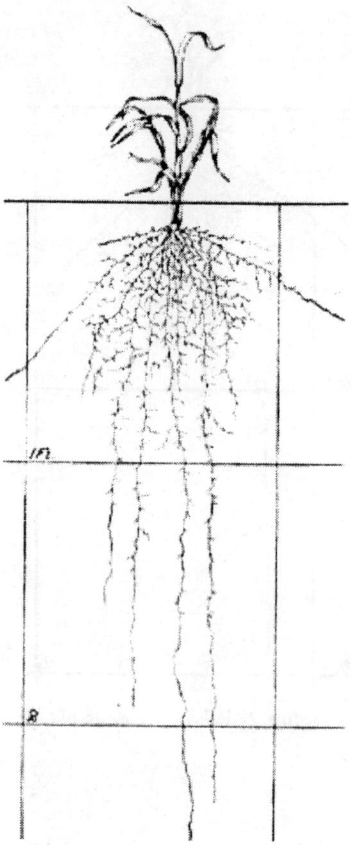

Figure 3. Wheat 60 days old.

The roots had deepened the working level to 1.5 feet, an increase in depth of about 10 inches.

Mature Root System

The plants were 2.2 feet tall, but the root system reached a maximum depth (4.8 feet) which was more than twice as great as the height of the shoot. During these last 20 days, there had occurred a marked development of roots (Fig. 4). The working level was at 3 feet; the lateral spread had increased to a maximum of 12 inches. A vast network of rebranched laterals occupied a volume of soil extending approximately 10 inches on all sides of the plant and to a depth of 2 to 3 feet. The total number of roots varied from 20 to 25 according to the number of tillers. Many of these were more superficially placed than those in the earlier stages of development, running rather horizontally or obliquely and ending in the surface 3 to 8 inches of soil.

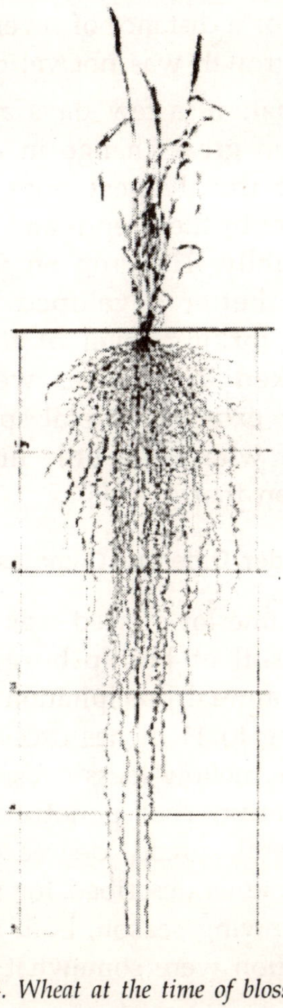

Figure 4. Wheat at the time of blossoming.

In the surface 2 feet especially, laterals were exceedingly abundant, usually 5 to 9 occurring on an inch of root length. Many of these were short, and few exceeded a length of 4 inches. Secondary laterals were not abundant. In the second foot, the branches were mostly less than 1 inch in length. Below 2 feet, branching was somewhat less pronounced, and in the fourth and fifth foot, numerous roots were characterized by unbranched, very short laterals. The fact that some root ends were without branches for a distance of several inches from the tips showed that growth was not yet complete.

A final examination, a few days after harvesting on July 15, showed no great change in root development. The roots, except the deepest ones, were somewhat shrivelled and more brittle. Depth and lateral spread had increased only slightly. The crop on the lowland was 4 inches taller and better developed than that on the upland and the working level of the roots 6 inches deeper. No marked differences were found in the branching habit or extent of lateral spread. As a whole, the root system of wheat is a little finer and somewhat more extensive than that of oats.

Root Variations under Different Climates

A crop from the same lot of seed was grown in mellow, fine sandy loam soil at Phillipsburg, in north central Kansas. Here, the annual precipitation is only 23 inches. But probably owing to 11 inches excess precipitation the preceding year, the mellow loess subsoil was quite moist beyond the maximum root penetration, 5.8 feet. The working level, lateral spread, degree of branching, etc., were about the same as described for Lincoln, in eastern Nebraska. The following season, both working level and maximum penetration were somewhat less.

Wheat was also grown in the hard, fine sandy loam soil of the short-grass plains at Burlington, Colo. Here, the 17 inches of annual precipitation moistens the soil to a depth seldom greater than 2.5 feet. Moreover, cold nights in early spring delay crop development, while later drought dwarfs the plants. Marquis spring wheat, grown on land that had been broken for 2 years, reached a height of only 1.7 feet, notwithstanding that the season was unusually favorable for crop growth.

The mature root system was confined entirely to the first 2.7 feet of soil, since no available water occurred deeper. Not infrequently, roots extended laterally 10 to 12 inches only 6 inches below the surface. The maximum lateral spread exceeded that of plants grown further east, and the entire root system was more profusely branched. Thus, the roots, although more shallow than normal, were well adapted to extract water and solutes from these surface soil layers of low water content. The marked difference in the degree of branching here and in eastern Nebraska is shown in Fig. 5. During the next season, these findings were confirmed, the root system being slightly less extensive.

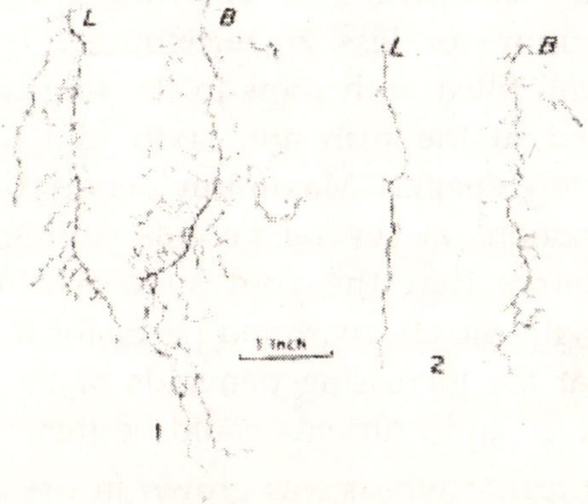

Figure 5. Wheat roots showing normal differences in branching at Lincoln, L; and Burlington, B; 1, at depth of 1.5 feet; 2, root ends.

The dry soil, watered by light showers, stimulated the development of an intricately branched and extensive surface-absorbing system at the expense of depth of penetration. In this widely spreading, surface-rooting habit, the crop behaved like the native grama and buffalo grasses. The tops were 2.5 feet tall. The roots extended to a similar depth, where dry soil prevented their further development. Lateral spread was marked and branching was profuse.

In the mellow loess soil along the Missouri River at Peru, Nebr., the shallower portion of the root system of Marquis wheat was not highly developed. But the part fitted to absorb in the deeper soil made a vigorous growth, having a working level of 4 feet. Many roots penetrated deeper, a few to 6.7 feet. Durum wheat (Triticum durum) was also grown at Peru. Compared to other cereals, it has a rather meager surface-feeding system at maturity. Usually, this consisted of six to eight (rarely more) roots that extended out in an almost horizontal direction from 2 to 14 inches. They usually ended only 4 to 7 inches below the soil surface.

The primary roots, accompanied by others, however, ran vertically downward or downward and outward, pursuing a more or less zigzag course. The soil was especially well filled with roots to the fourth foot, many also occurred in the fifth and sixth foot, and several extended even deeper. Maximum penetration was 7.4 feet. Examinations at several periods during its growth showed clearly that the root system developed in correlation with the aboveground parts, for it was only in this way that the increasing demands of the developing shoot for water and nutrients could be met.

Marquis spring wheat was grown in dry land and in irrigated soil at Greeley, Colo. Since the precipitation is only 13 inches annually, irrigation is widely practiced. The fine sandy loam soil in the several plots was of very similar physical and chemical composition. The wheat plots were treated alike as regards seed-bed preparation, time and rate of seeding, etc., except that the irrigated plots had been fertilized uniformly with 5 tons of barnyard manure per acre.

Early Development

Root development on May 10, when the crop was about 6 weeks I old and in the fourth leaf stage, is shown in Fig. 6.

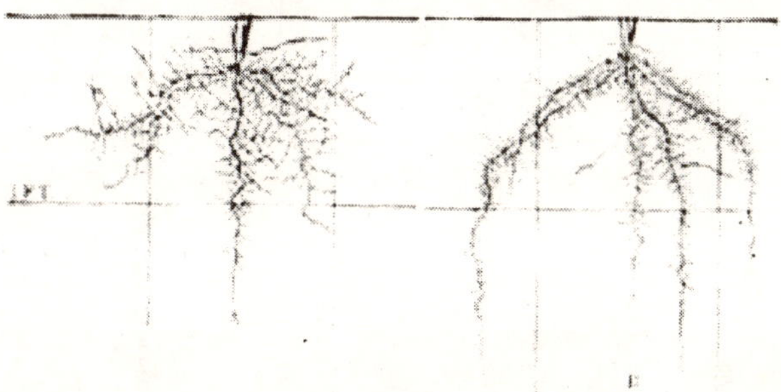

Figure 6. Root system of Marquis spring wheat 6 weeks old: A, in dry land; B, under irrigation.

The roots in irrigated soil developed quite normally. In the dry land, the available water in the second foot of soil was almost exhausted. Hence, the roots were mostly confined in their distribution to the surface 12-inch layer. The number of roots and branches was about the same in both cases, but the branches averaged considerably longer in the dry land.

Half-grown Plants

A month later, although a small amount of water was

available in the second foot of soil, the dry-land crop was only 13 inches high and showed distinct signs of suffering from drought. The plants had only one or two tillers each. The irrigated plants were 21 inches tall and had about twice as many tillers. Differences in root habit were very striking (Fig. 7).

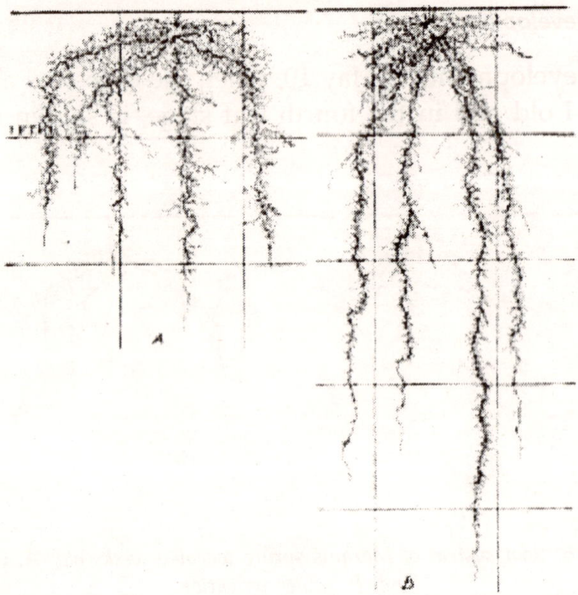

Figure 7. Roots of wheat plants 2.5 months old: A, in dry land; B, in irrigated soil.

The wider spread, longer primary branches, and the much greater number of secondary and tertiary laterals in the drier soil are clearly evident. The network of roots just beneath the soil surface afforded an efficient means of securing water furnished by light showers. But many of these roots had died from drought, and growth was greatly retarded. Under irrigation, lateral spread was much less developed, but the root system extended very much deeper. Maximum penetration in the two cases was 31 and 65 inches, respectively.

Mature Root Systems

A final examination was made when the crop was nearly ripe. The very meager rainfall during the interval since the last examination was entirely dissipated in several light showers, and the soil in the dry land had become progressively drier. Here, the wheat was only 15 inches tall.

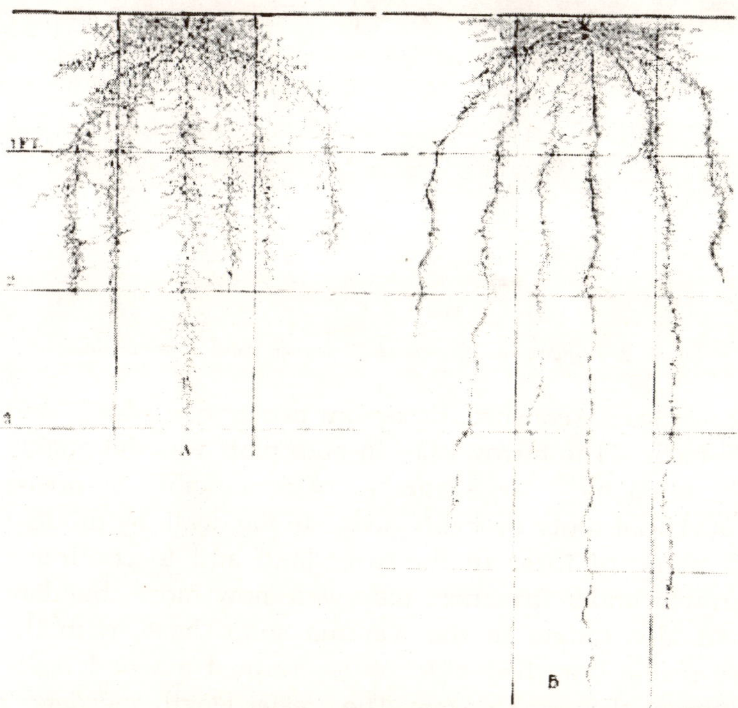

Figure 8. Roots of mature wheat plants: A, in dry land; B, in irrigated soil.

Only about half of the plants were furnished with a tiller, very few of which had headed. The yield was scarcely 3 bushels per acre. A fine crop, 43 inches tall and yielding at the rate of 29 bushels per acre, had developed under irrigation. Differences in root habit were quite as marked as before. A comparison of Figs. 7 and 8 shows that the roots in the dry land had grown relatively little. The chief differences were a more thorough occupation of the second and, to a slight extent, the third foot of soil. The working depth was 24 inches as compared with 52 in the irrigated plots.

Figure 9. Primary root system of 10-day-old plant of winter wheat.

In the same sequence, maximum penetration was 37 and 75 inches. The furrow slice in both plots was thoroughly occupied by a large number of remarkably branched superficial roots. Probably owing to the death by drought of many of these in the drier land and to continued growth under irrigation, they were now more abundant and also longer in the watered soil. These, with the profusely branched older roots, formed a wonderfully efficient absorbing system. The greater length and degree of branching of laterals were, as before, very conspicuous in the drier soils.

Variations in Root Development

The following season, when the soils were equally moist, due to increased rainfall, the first examination revealed no differences in root habit. Later, the area occupied by the root system of the dry-land crop was much greater than during the preceding season, owing to a better shoot development, more tillers, and a subsoil with available moisture in which none of the roots died. The lateral spread was as great as formerly, and the working depth was over a foot deeper. In fact, the root habit was more nearly like that in the irrigated soil than that in dry land. The crop was 3 feet high and the yield 25 bushels per acre. Root development in the irrigated soil was approximately the same as the preceding year. A third plot, where light irrigation was practiced both years, gave results intermediate to those just described.

Winter Wheat

Development of winter wheat under measured environmental conditions has been thoroughly studied at Lincoln, Nebr. A strain of Turkey Red winter wheat (Triticum aestivum), known as Kanred, was grown. It was drilled 2 inches deep in fertile silt loam soil on Sept. 20, and the growth both above and below ground recorded at 10- or 15-day intervals. Growth conditions were very favorable during both years of the experiment, and the crop developed normally.

Early Development

Ten days after sowing, when the second leaf was about half grown, the roots were excavated (Fig. 9). The number of roots varied from two to five, but nearly all of the plants had three. The primary roots were deepest, extending to maximum depths of 8 to 9 inches. While these roots took a rather vertically downward course, the others usually ran obliquely outward, often later turning downward. The fairly abundant supply of laterals was scattered quite irregularly, the best-branched portions of the root giving rise to 12 or more per inch.

Ten days later, the plants had four leaves, and nearly all had one tiller extending an inch or more from the axil of the first leaf. The leaves were rapidly losing their vertical position, some already being nearly horizontal. A second tiller, originating either from the axil of the second leaf or more commonly below ground near the grain, was also found on most plants. Nearly every plant had a new root in addition to those of the primary root system. These roots of the secondary system were about a millimeter thick, turgid, white, and entirely unbranched but densely clothed with root hairs.

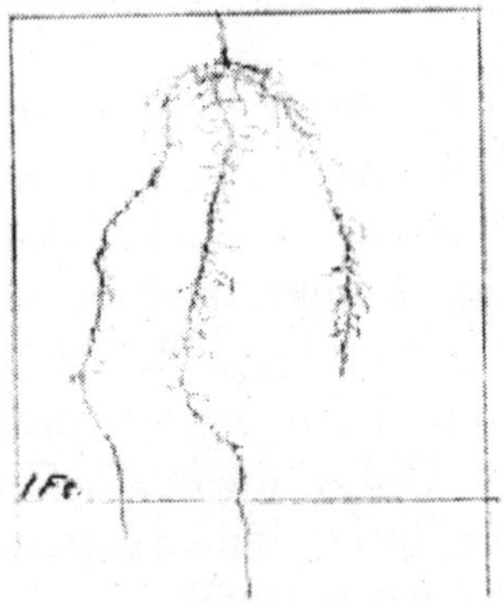

Figure 10. Wheat plant 20 days old. The first root of the secondary system has appeared.

They originated from the stem near the grain and ran mostly in a horizontal position or turned only a little downward. None exceeded 2.5 inches in length. The roots of the primary system had extended into the second foot of soil, elongating at the rate of over half an inch per day (Fig. 10). They were more frequently branched, and the branches were longer than before, but no laterals of the second order had appeared.

When 30 days old (Oct. 20), most of the parent plants had the fifth leaf about half developed. Each was furnished on an average with four tillers. The largest of these had three leaves and stood as high as the parent plant. The prostrate habit, due to the outward curving of the short stems, was well initiated, the plants having a spread of 3.5 inches on each side of the drill row. This rapid growth of tops was correlated with a pronounced root development. The primary root system had reached a working level of 16 inches and a maximum penetration of 2.8 feet was attained by some plants (Fig. 11). The long, thick, unbranched root ends indicated rapid growth. Not only was the lateral spread greater, but the branches were much longer, and the older portions of the roots possessed many more of them. Branohes of the second order were found only on the oldest laterals from the main roots, but here they were abundant.

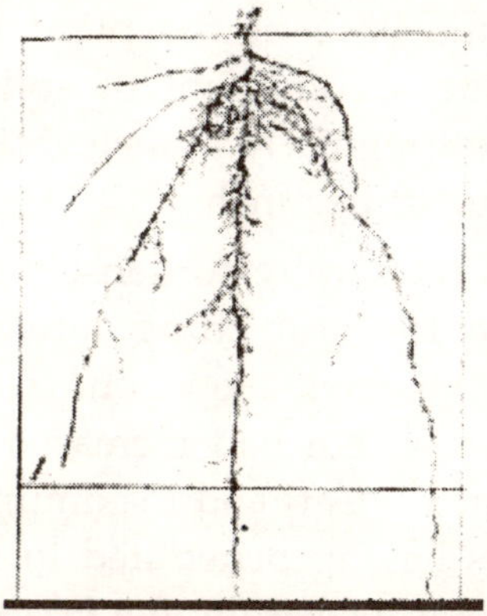

Figure 11. Wheat plant 30 days old. The secondary root system now furnished 18 per cent of the absorping area.

The secondary root system, moreover, had made a marked growth. Each plant now had a total of 4 to 10 roots, an average of 4 in addition to the seminal ones. They varied from 1/8 inch to 6 inches in length and turned downward from horizontal to nearly vertical. They had about twice the diameter (1 millimeter) of the seminal roots in the surface soil, and only in the deeper layers was the diameter of the latter equal to that of the roots of the secondary system. All were densely clothed with root hairs, and short laterals occurred on the older roots.

The next 10-day period revealed a marked growth. The number of tillers had increased to 7 per plant. These were so well developed that a plant of average size had a total of 20 leaves, more than half of which were fully grown. To provide water and nutrients for such a large shoot, an extensive root system was imperative. An examination of the latter showed that the roots of the secondary root system averaged 9 per plant. While some were only a small fraction of an inch long, others had a length of 19 inches. In general, they ran rather obliquely outward and downward with an average spread of about 5 inches from the base of the plant (Fig. 12). A few extended into the second foot of soil. The older and longer ones were irregularly branched with short laterals at the rate of 5 to 10 per inch.

All the main roots were so densely clothed with root hairs, to, which the soil clung tenaciously, that the smooth, white, root ends .stood out in marked contrast. The primary root system had increased both in working depth (now about 1.7 feet) and maximum extent, a few of the deepest roots having penetrated to 3 feet. The oldest portions of the roots, especially the first foot of the deeper ones, appeared shriveled, and microscopic examination showed a deterioration of the cortex. But the abundant root hairs on the deeper main roots and their branches, together with their bright, turgid appearance, showed plainly that they were functioning vigorously as absorbing organs.

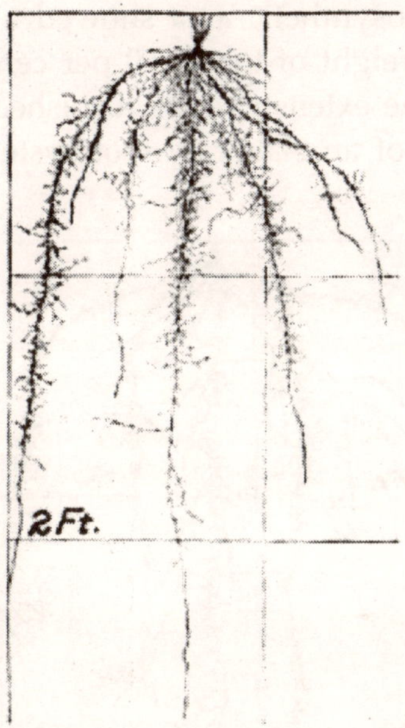

Figure 12. Root system of 40-day-old wheat plant.

Primary branches were longer than before, and on some, secondary laterals were much more abundant. Thus, the efficiency of this portion of the root system was greatly increased.

Late Autumn Development

After a 15-day interval, further examinations were made, the crop now being 55 days old. Growth had been very good so that, although the drill rows were 8 inches apart, over much of the field the soil was practically concealed by the plants. The height was only 3 inches, owing to strong curving of the short stems which gave the plants their favorable prostrate winter habit. Tillers had increased rapidly from 7 per plant, 15 days earlier, to 11. On an average, each plant now had 32 leaves, an increase of 11. The photosynthetic area showed a gain of 141 per cent, and dry weight of tops, 160 per cent during the 15-day interval. The extensive tops furnished abundant food for the growth of an elaborate root system.

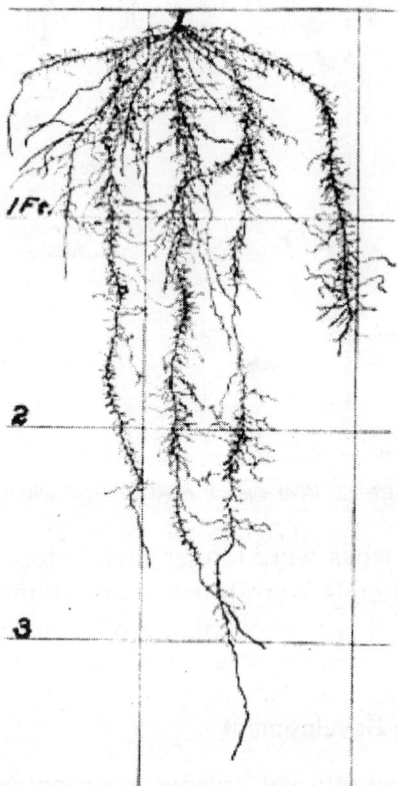

Figure 13. Root system of 55-day-old wheat plant.

Roots of the primary system often reached depths of over 3 feet, and a few were found at the 4-foot level (Fig. 13). Branching habit of the primary system had changed but little in the surface 2 feet except that the branches were somewhat longer. The younger portions, as the roots deepened, became clothed with laterals similar in number and in secondary branching to the older parts above.

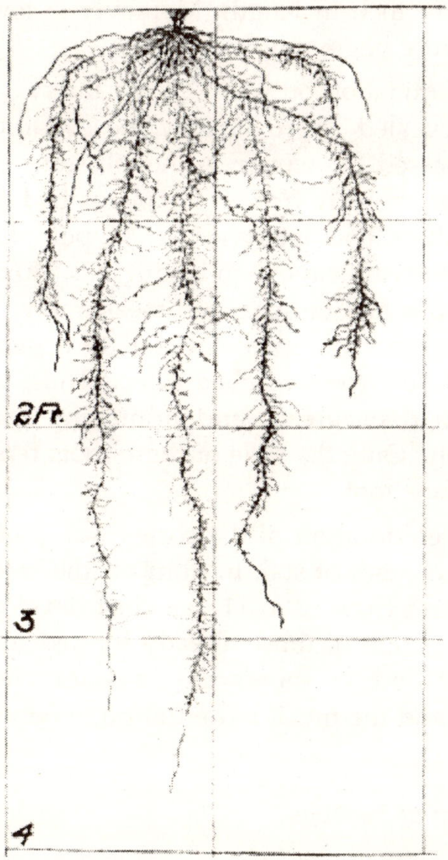

Figure 14. Root system of wheat 70 days old, showing the extent of root growth before the period of winter dormancy.

An average of 10 roots of the secondary system was found. They ran almost horizontally or so slightly obliquely downward that few or none occupied the area under the plant where the roots of the primary root system were absorbing. The working level was at 8 inches, but some of the longer roots penetrated to a depth of 20 inches. About half were unbranched or nearly so; others were profusely branched throughout with laterals averaging an inch in length. Moreover, a few secondary branches were beginning to appear.

An examination on Nov. 29, the interval again being 15 days, revealed approximately the condition in which the roots passed the winter. Although some of the tips of the older leaves were frozen, the plants had made a good growth. Tillers had increased to 14 per plant, and the total number of leaves, to about 40. The height and spread of tops had not changed measurably. The primary root system now occupied the soil to a depth of 3 feet (Fig. 14). A few roots reached 4 feet. Thus, some growth in depth had occurred, and branching had increased considerably. Only the stele of these roots remained intact in the surface foot.

Early deterioration of the cortex was probably due to low water content of soil. In number, the secondary roots had increased from 10 to 11, the chief development being in the elongation of those already formed. A comparison of Figs. 13 and 14 shows the considerable increase in branching and the much more thorough occupation of the soil.

Development of the Tops

The relative growth of the tops, including photosynthetic area and dry weight, at the end of the several intervals. The measurements throughout the period of winter dormancy until growth was resumed in the spring. The tiller production started about 15 days after planting and was kept up continuously until the middle of December. Leaf output paralleled the growth of tillers, and growth rate, based on dry weight, was very similar. The primary root system increased its working level quite uniformly. This was at the rate of 0.55 inch per day during the first 55 days, the extent of branching correlating with root elongation. The secondary root system began to develop simultaneously with the appearance of tillers.

On an average, a new root and a new tiller appeared every 4 or 5 days until the middle of November, after which the rate of tillering exceeded that of root production. However, the increase in length and branching of the secondary root system continued with the formation and growth of tillers and at an undiminished rate until the middle of December. Here, growth both above and below ground ended abruptly when the air temperatures averaged almost continuously below freezing and the soil was frozen to a depth of several inches. The number of leaves and tillers decreased slightly due to repeated freezing and thawing, wind whipping, etc., and consequently, the dry weight of tops also decreased. The greater decrease in the latter (44 per cent) was due to the fact that many of the leaves were only partly injured, causing a marked decrease in dry weight but not in numbers. But the root system, even that part subjected to the greatest temperature changes, was apparently uninjured.

During the second week in March, both roots and shoots resumed growth. Frost disappeared from the soil as the air temperature became higher, and the plants developed slowly at an average temperature of 40°F. Rains replenished the water content of the surface soil, and with increasing temperature, the crop made a steady growth. The primary root system was apparently functioning as in late fall. Many new roots of the secondary system appeared. The culms began to grow erect and the spikes to develop. In less than 2 weeks, at an average temperature of 42°F., the crop regained 60 per cent of the loss in dry weight which had occurred during the 90 days of winter dormancy. This vigorous early spring growth was due largely to the well-developed root system and culms stored with food.

Absorbing Areas of Roots

Determinations were also made of the relation of the actual absorbing area (exclusive of root hairs) to that of the transpiring area of leaves and stems. The work was done during the following season when conditions for growth were very similar to those described. This was accomplished by carefully washing the soil from the roots with a gentle water spray, thus securing the root systems in their entirety and floating them in shallow trays of water while measurements were being made.

The absorbing area of roots increased progressively with that of tops and was uniformly 10 to 35 per cent greater in extent. Since microscopic examination indicated that practically all of the roots and their branches were clothed with functioning root hairs, except at the growing tips and on the oldest parts of the primary roots, the absorbing area was actually probably eight to ten times greater than the transpiring area. Deterioration of the cortex on the oldest portions of the roots of the primary root system began about the middle of November.

By the end of the month, the epidermis on about the first 10 inches was either destitute of root hairs or sloughed off with the cortex, leaving only the stele intact. This reduced the absorbing root area, however, by less than 1 per cent. On Oct. 20, the secondary root system already furnished 18 per cent of the total absorbing area. This had increased to 40 per cent 10 days later, and by the middle of November, it was 51 per cent, notwithstanding the great increase of the area of the primary root system. Owing to the relatively finer roots of the primary system coupled with more profuse branching, its total length exceeded that of the secondary root system.

Mature Root System

At maturity, winter wheat has a very extensive root system. As with other cereals, the abundance of roots, lateral spread, and amount and length of branching, as well as the depth of penetration, are quite variable in different kinds of soil and under different climates. A representative specimen of the Turkey Red variety is shown in Fig. 15. It was grown in moist, rich, silt loam soil near Lincoln. The tops were 3.8 feet high and the heads were well filled. Most of the numerous thread-like roots penetrated rather vertically downward, others ran obliquely downward but seldom reached a greater spread than 6 to 8 inches from the base of the plant. Still others ran out parallel with the soil surface for short distances before turning downward. The working depth was found at approximately 4.4 feet, and the maximum root depth was 6.2 feet.

Beginning just below the surface and extending to a depth of 4 feet, numerous profusely branched laterals filled the soil. These light-coloured roots showed very plainly in the black earth. They were covered with dense mats of root hairs, the rootlets intercrossing in the jointed subsoil in such a manner as to give a cobwebby appearance.

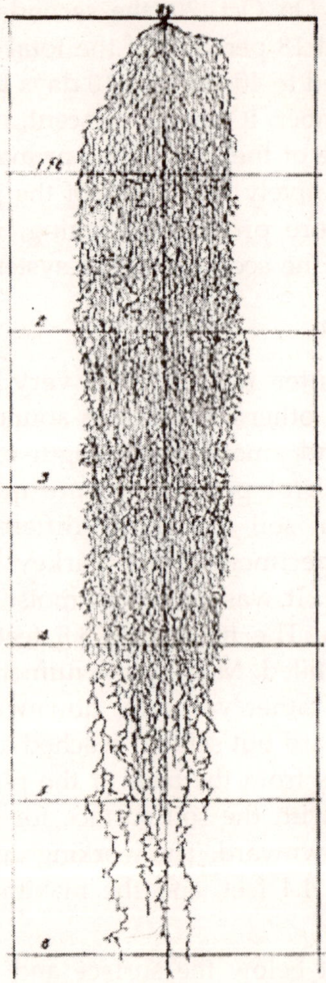

Figure 15. Mature root system of winter wheat.

It is quite impossible to show these finer roots and all their branches in the most detailed drawing. Below 4 feet, the roots were less abundant but still well branched and supplied with root hairs. The last 6 inches of the deeper ones were poorly branched with laterals which were only a few millimetres in length.

Root Variations under Different Soils and Climates

A field only 2 miles distant from that last mentioned, where silt loam intergraded at a depth of 2.3 feet into a very hard tenacious subsoil of clay intermixed with streaks and spots of chalk, gave marked differences in root extent. The crop was 3.3 feet high and of excellent quality. The working depth was only 3.2 feet and the maximum root penetration 4.7 feet, depths approximately 12 to 18 inches less than in the deep silt loam soil. However, in a third field, an equal distance from the first and also examined the same season, much greater root extent was found. Here, the silt, loam soil gradually gave way at a depth of about 1.5 feet to a very deep, rather mellow, loess subsoil. Like that in the other fields, it was moist to great depths. The mature crop was 3.5 feet high. The lateral spread was similar to that described, but roots were fairly abundant to the working level at 4.9 feet. Not a few penetrated to 7 feet, and a maximum depth of 7.3 feet was attained.

Quite in contrast to this excellent growth was that on the short-grass plains. At Flagler, Colo., a field of the Turkey Red variety was rooted entirely in the surface 16 inches of soil. The roots were developed very much as if grown in a large flowerpot, because the soil was moist only to a depth of 15 inches where a very tenacious hardpan, 7 inches thick, occurred. Below this, the soil was less compact but very dry. Such a limited root development was correlated with a poor growth of tops which scarcely exceeded a foot in height.

12

Red Clover

Red clover (*Trifolium pratense*) is a plant belonging to the pea family Fabaceae. It is often used as fodder. The plant was named *Trifolium pratense* by Carolus Linnaeus in 1753. Trifolium means 'three leaved' while pratense is Latin for "found in meadows".

Red clover flower

Isoflavones from red clover have been used to treat the symptoms of menopause. It has also been reported that red clover can be used for therapeutic purposes for coughs, bronchitis, eczema, sores, scrofula and can be gargled for mouth ulcers and sore throats. Women who are pregnant and breast-feeding should avoid ingesting Red clover.

A perennial with trifoliate leaves and pink to red flowers. Red clover is often planted as a component of pasture and forage mixes, but sometimes escapes to become a weed of turfgrass, lawns, landscapes, and orchards. Red clover is distributed throughout the United States.

Cotyledons are spatula-shaped, 6-7 mm long, without hairs. The first true leaf is solitary, oval in shape, and cut off squarely at the base (truncate). All subsequent leaves consist of 3 leaflets (trifoliate). A fibrous root system and stems that root at the nodes.

Red clover consist of 3 leaflets (trifoliate). Each leaflet is elliptic in outline, approximately 1/2 to 1 1/4 inches long and about half as wide. Leaves are usually hairy on both surfaces or sometimes without hairs above and hairy beneath. Each leaf usually has a light green or white 'V-shaped' marking.

Growing somewhat prostrate but ranging from 8 to 20 inches in height. Stems are sometimes without hairs but can be slightly hairy, and root at the nodes. F l o w e r s occur in round to oval heads usually on flower stalks (peduncles) less than 5 mm long, but also without flower stalks (sessile). Individual flower heads are pink to red in colour, 1/2 to 1 1/4 inches in length. Fruit: A pod known as a legume, 4 to 5 mm long.

Red Clover Cultivars with Grasses

Red clover is most often grown with companion grass for hay, silage or grazing. However, little information is available on the performance of cultivars in competition with grass species. Smith and Sharpee suggested that the relative performance of red clover cultivars may not be similar when grown in monoculture vs. in competition with grass.

As expected, the yield of clover grown without grass was, in general, greater than when grown with grass, although the total grass-clover yields was greater. The performance of red clover cultivars was influenced by the location, the cutting system, and the presence or absence of grass. In general, the performance of the cultivars grown without grass did not predict the corresponding response when grown with grass. Of the 15 possible comparisons of the performance of the cultivars with and without grass only one was significant.

In fact, this significant correlation, -0.64, was in the opposite direction than expected. If performance in pure stands (without grass) is going to predict performance with grass the correlation should be positive and high (greater that 0.62 but less than 1.00). Future selection of red clover cultivars should be done in the management system that are similar to the system used by the growers.

It has been long recognised that red clover is an excellent forage legume for hay, silage, and grazing. As indicated earlier, the use of red clover in the US has declined over the past 100 years from about 30 million acres in the early 1900s to about 11 million by 1987. This decline in use has primarily been due to low nitrogen fertiliser prices, an increase use of alfalfa, a decrease in legume-grain row crop rotations, and the short-life of the plants. However, since the 1980s the use of red clover has stabilised, if not increased somewhat.

This has been due to the development of improved, persistent, disease resistant cultivars; the recognition of its value in grazing conditions; and an increased recognition of its importance in crop rotations and soil conservation. Recent research has identified chemical compounds in red clover, some of which are important in protein utilisation, others in animal reproductive processes, and still others as pharmaceutical products. Therefore, the use of red clover will continue to occupy a position of major importance in US agriculture.

Improvement in Agriculture

Disease resistance and persistence: Even though current cultivars have good disease resistance and are relatively persistent, continued improvement in these attributes is needed. Lack of resistance to the soil-borne Fusarium pathogens is one of the leading causes of stand failure in red clover. Some improvement has been made in this direction, but further, concentrated efforts are needed and should be rewarded with even a greater degree of persistence in the crop.

Crown position and root structure: Unlike alfalfa and some other forage legumes where the crown of the growing plant is positioned just below the soil surface, the crown of red clover is positioned at the soil surface. This exposes it to a greater degree of environmental stresses such as temperature, disease organisms, physical abuse, etc. Developing red clover cultivars where the crown is position just below the soil surface would give greater protection against these stresses and undoubtedly lead to increased persistence.

Initially red clover produces a primary tap root, but as it is exposed to diseases and other stresses it gives way to a secondary, branching system that develops reasonably close to the soil surface. Types need to be identified that have a disease tolerant tap root structure with an extensive secondary system. Rhizomes, underground stems, have not been observed in the red clover species, however, related species do have such structures. Efforts need to be extended to interspecifically transfer, through molecular or other techniques, this desirable attribute. Incorporation of the rhizome structure would enhance persistence and improve drought tolerance in the species.

Grazing tolerance and grass competition: Red clover is being used extensively for grazing and most often is grown with a companion grass whether grazed or used for hay or silage. Very little effort has been expended to develop red clover cultivars that are adapted to these conditions. The red clover cultivars in monoculture is not indicative of their performance when grown with grass companion crops. It is presumed that this would be true whether used for grazing or otherwise. A major component of any future red clover genetic and breeding projects should include the development of germplasm tolerant to these uses.

Quality Improvement

Protein utilisation and protection: Papadopoulos and McKersie reported that red clover has 90% less proteolysis than alfalfa during ensiling. Also, Albrecht and Muck report ensiled red clover yields 30 to 40 % less non protein nitrogen (NPN) than ensiled alfalfa. If proteolysis and NPN formation can be reduced, then forage protein utilisation by the ruminant animal should be improved. Jones *et al.* report that this effect in red clover is the result of the polyphenol oxidase (PPO) enzyme system in red clover that inhibits activity of the plant's proteases in the silo. Among the forage legumes this PPO enzyme system is unique to red clover. This important attribute in red clover should be exploited by either genetically increasing its activity in red clover or genetically transferring the attribute to alfalfa and other legumes or both.

Phytoestrogens: Red clover contains high levels of phytoestrogens which have been reported to interfere with normal reproduction in sheep. Although no documented reports have been recorded for dairy cows it remains that this may present a problem, especially for grazing cows. The oestrogenic compounds are isoflavones which include formononetin, diadzein, genistein, and biochanin A. Formononetin (7-hydroxy-4′-methoxyisoflavone) is recognised as the most important of these oestrogenic compounds in red clover.

Selection for lower levels of formononetin in red clover has been effective. The improved fertility of ewes grazing the low formononetin red clover. Since red clover is being used extensively in grazing conditions, red clover cultivars with low levels formononetin need to be developed. Contrastingly, dietary compounds, Promensil, with high estrogenic activity are being market as a supplement in human diets to combat PMS and osteoporosis.

Pectic polysaccharides and forage quality: While protein content in most forage legumes is quite high, the full potential of the protein, especially the leaves, is not realised due to rapid degradation rates and conversion to ammonia in the rumen. Numerous schemes to alter the proteins of forage legumes range from chemical to physical treatments to actually genetically altering the plant to produce proteins more resistant to degradation. An alternate strategy to improve utilisation of plant protein would be to match the rapidly degraded protein with an equally rapidly degraded carbohydrate (RDC) source such that energy would not be limiting to rumen microbes.

This assures that the plant protein would be efficiently converted to microbial biomass benefiting the ruminant. Pectic polysaccharides (commonly known as pectins) represent a potential source of RDC , and a significant portion of the total structural polysaccharides in forage legume cell walls are pectic materials. Thus, increasing the cell wall pectic polysaccharide (pectin) content of forage legumes would improve utilisation of their nutritional potential. Earlier research has shown that genetic variability exists in red clover for cell wall pectin concentration. Divergent populations of red clover and other forage legume germplasm with a range of cell wall pectin concentration are needed to assess the actual impact of increased plant pectin on forage quality and animal performance.

Medium Red Clover

Medium Red Clover is the most widely grown of the true clovers. It is a short-lived perennial legume native to the countries that border the Mediterranean and Red Seas. Medium Red Clover is used for short rotation hay fields and include into pasture mixes with orchardgrass & timothy or tall fescue. Medium Red Clover is a perennial which acts as a biennial under usual farm conditions. The clover root border and crown and root rots kill many plants after the second cutting of the first hay year. If red clover is left over the second winter, winter injury, combined with diseases and clover root borer, further deplete the stand.

Red Clover normally produces two cuttings during the hay year. Red Clover grows on soils with pH values below those necessary for satisfactory production of alfalfa and sweet clover. It is easier to establish in wheat or other winter grains than alfalfa. Red Clover may be seeded in oats. It is not as well adapted to summer seeding as is alfalfa; however, early August seedings frequently are successful. It is often noted that when red clover blooms during the fall of the seeding year, many plants die during the winter.

Blooming causes a physiological change in a red clover plant which makes it less winter hardy. In addition, excess summer and fall growth which is not removed from the field, will lodge and create conditions favourable for the development of crown and root rots. These diseases can damage new seedings. The general rule-of-thumb is that new seedings of red clover which have reached a height of 8 to 10 inches by mid-August should be harvested no later than September 1, regardless of the presence or absence of bloom.

Seeding Rate

Red clover (Trifolium pratense L), a short-lived perennial, is the most widely grown of all the true clovers. Classifying red clover is sometimes confusing, but the red clovers grown in the United States may be grouped into two divisions—early flowering and late flowering. The more useful of the two types grown in Missouri is the early-flowering type, usually referred to as medium red clover. This type produces two or three hay crops per year and is usually treated as a biennial although it is actually a short-lived perennial. Some new medium varieties will produce to their full capacity for three years or more.

The late-flowering types are referred to as mammoth red clovers. They usually produce one hay crop plus an aftermath growth. Mammoth red clover has yielded from 15 to 40 percent less hay in tests throughout the lower Corn Belt. Mammoth red clover is not recommended for use in Missouri; its best performance is in areas with short growing seasons such as Canada.

Use and Adaptation

Red clover is effective as a pasture legume and for hay. It can withstand more shading in the seedling stage than most other legumes, making it easy to establish in grass sods. To make red clover more effective, follow these management practices:

- — Maintain fertility level—test the soil and topdress accordingly.
- — Use rotational grazing, especially during the second year.
- — Rest the pasture in September and early October.

Medium red clover is most productive on soils of medium to high fertility levels with a soil salt pH of 5.5 or higher with good internal drainage. It is better than alfalfa at tolerating and growing on soils of low pH and low fertility with poor drainage.

Selecting Seed and Varieties

Always use medium red clover rather than mammoth. Two varieties of medium red clover—Kenland and Kenstar—are superior to others in Missouri. Buy certified Kenland or Kenstar seed to guarantee it is not contaminated with mammoth red clover or with varieties not adapted to Missouri. Use resistant varieties to control diseases that can seriously damage red clover, such as northern and southern anthracnose, powdery mildew, and root and crown rots.

Kenland

Kenland red clover was released several years ago. It has good resistance to southern anthracnose, which is especially destructive to red clover stands in the southern Corn Belt. It has superior yielding ability over most other varieties and has a longer life than common red clover. It reportedly has more resistance to crown and root rots than seed originating from the northern United States.

Kenstar

Kenstar is a new variety of medium red clover released by the University of Kentucky. It is very similar to Kenland in its area of adaptation and its resistance to southern anthracnose and powdery mildew. It is slightly superior to Kenland in yielding ability during its first and second harvest years.

Kenstar's greatest advantage over Kenland is its ability to persist and produce during the third harvest season. When sufficient seed is available, Kenstar is expected to replace Kenland for general use in Missouri. Kenstar is protected under the Plant Variety Protection Act, and use of the name Kenstar is prohibited beyond the certified seed class.

Common red clover

This refers to all classes of seed of unknown varietal origin. Unless a variety is certified, it might not be genetically pure. One of the great concerns about using mammoth clover, aside from the yield reduction to the first grower, is the propagation of this inferior strain by seed harvest. It might find its way back to Missouri farms as common red clover seed.

Although there is a greater advantage to using certified red clover seed compared to other commonly grown forages, many pounds of common red clover are seeded in Missouri each year. If you must use common seed, locally produced seed (although there is no way to tell its genetic purity) is usually superior to seed produced outside the southern Corn Belt. If you purchase from an individual or neighbour, have a sample tested for germination and for the presence of troublesome weeds. Avoid seed low in germination that contains a high percentage of troublesome weed seeds.

Table 1.

Red clover yields	1974	1975	Total tons
Kenstar	4.85	2.93	7.78
Kenland	4.06	2.46	6.52

When buying common seed from dealers, check the tag to verify its purity and origin. Red clover seed sold under a brand name with no specified variety has no guarantee as to its genetic purity.

Seeding Red Clover

You can seed red clover in late summer or early fall. This usually increases total yield for red clover during its lifetime as compared with spring or winter seedings. One of the most common ways to establish red clover is to broadcast the seed on winter wheat fields during late winter. Freezing and thawing will work the seeds into the soil enough for germination to occur. One of the drawbacks of this method is that good clover stands may be difficult to obtain in heavy wheat stands that are harvested for grain. To avoid smothering the clover, clip the straw and remove it as soon as possible after the wheat harvest. Shredded straw can also aggravate seedling diseases.

You can also seed red clover with spring oats. Red clover is well adapted to sowing on a prepared seedbed, either in the early spring or late summer (Aug. 15 to Sept. 14). The seedbed should be firm, and the clover seed should not be covered with more than 1/4 to 1/2 inch of soil.

Fertility treatments are best recommended on the basis of a soil test. In the absence of a soil test, use a fertiliser treatment of about 20 pounds of nitrogen, 60 pounds of phosphate and 40 pounds of potash at seeding time when you are seeding on a prepared seedbed in the fall.

When red clover is seeded on small grains during winter, some phosphorus (about 60 pounds) topdressed at the time of seeding has proven beneficial, especially if the soil is low in phosphorus. You can also mix seed and fertiliser together and apply them at the same time.

Seeding Rates and Mixtures

Red clover is best grown with a grass. The stand will remain longer and will continue to be productive when the legume has thinned. Also, the danger of bloat with red clover is greatly minimised when it is grown with a grass. All of the cool season grasses adapted to Missouri can be used with red clover. Timothy has universally been used, and it is an excellent choice. It is a quality grass, and it matures at about the same time as red clover.

Usually about 8 pounds of medium red clover seed per acre is used in association with a grass. Examples of red clover-grass mixtures:

- — Red clover, 8 pounds; orchardgrass, 6 pounds.
- — Red clover, 8 pounds; timothy, 2 pounds (fall) or 4 pounds (spring).

— Red clover, 8 pounds; tall fescue, 10 pounds.

— Red clover, 8 pounds; reed canarygrass, 6 pounds.

Management in year of Establishment

Red clover seeded in winter or early spring may be grazed the first summer. If it blooms before Sept. 10, it should be removed. The most practical way to prevent this is to graze during August. Do not graze from Sept. 10 to Nov. 1 or frost, whichever occurs first. Root reserves are important for building up winter hardiness and disease resistance.

Usually some fertiliser applied in fall or winter will benefit a new stand of red clover. This is especially true if the crop was not adequately fertilised at seeding time. Potash does not stimulate higher yields in red clover to the same degree as it does in alfalfa, but some potash is beneficial in maintaining stands. In the absence of a soil test, apply a mixed fertiliser such as a 0-30-90 each year to red clover fields used for hay.

Management in Second and Third Years

Grazing red clover in pre-bloom stage doesn't harm the crop. Because red clover is high in magnesium, it can be used for early spring grazing in the fight against grass tetany. Practice rotation grazing to get the highest yields and stand persistence into the third year. Removing the forage with three grazings will result in better third-year stands than four grazing harvests. If used for hay or hay and seed, two cuts or harvests are desirable. If possible, always rest the field in September and most of October. The fall growth need not be wasted. It can be grazed after Oct. 25 or frost, whichever occurs first.

Red Clover for Pasture Renovation

Red clover is an excellent choice for seeding in grass sods.

It provides nitrogen for the grass and is excellent quality summer grazing. Red clover seeded into fescue sods has increased yields about 35 percent above those plots receiving 120 pounds of nitrogen per acre.

Red clover is one of the easiest legumes to re-establish in predominantly grass sods. This may be done in one of several ways, but the grass must be controlled until new clover seedlings become established. High seeding rates (10 to 12 pounds per acre) have resulted in better stands than lighter seeding rates.

Method 1

Use a disc or field cultivator to disturb the grass sod from November through January or early February. Broadcast the clover seed early enough during winter so that freezing or thawing will cover the seed. Seedings made in mid-February have a 50 percent better chance of success than those made in mid-April. When early grass growth begins, graze or clip the new growth. Fertiliser containing phosphorus and/or potash may be broadcast prior to seeding. Don't use nitrogen in the fertiliser mix because it will stimulate the well-rooted grass more than it will benefit the young clover seedlings.

Method 2

This method is similar to method one, but the sod is not tilled. For this method to be successful, several management details are necessary. The grass should be weakened as much as possible by grazing prior to seeding. The seeding should be made no later than Feb. 15 in the northern part of the state and before Feb. 1 in the southern part. New spring growth should be grazed early in the growing season. This step cannot be overemphasised because early, repeated heavy grazing is a must to control the grass growth and retard its production while the young clover plants are establishing their root systems. It is best to use large numbers of cattle and to overgraze the grass quickly. This procedure should be repeated as necessary throughout late March (in South Missouri), April and early May.

Nitrogen fertiliser should not be used; it is best if the grass was not topdressed with nitrogen during the previous late summer or fall period. Adequate phosphorus and potash should be applied prior to seeding.

Method 3

This method eliminates a number of tillage operations, reduces soil erosion and can be done later in the spring than methods one or two. This method uses a chemical to retard grass growth. In most cases the chemical used has been paraquat. Since the chemical is a contact killer, seeding must be done after the existing grass has started its growth. Thus, it must be done either in early spring or late summer. Rates of one to two pints of paraquat has proven satisfactory.

Some type of grassland drill or seeder must be used to incorporate the seed into the undisturbed sod. Seedings made during the early spring after grass growth has started cannot be broadcast onto the soil surface with any great chance of success for a new seeding.

Seed Production

Missouri farmers have the opportunity to produce both certified and uncertified red clover seed. Seed harvest is best taken from the second growth of red clover in the year following the year of seeding. Highest seed yields usually are obtained from red clover when the first growth is used for hay. Hay should be harvested no later than 15 days after the earliest heads appear; this is normally early full bloom. Seed yields average about 1 or 1-1/2 but can be doubled or tripled with proper cutting management if sufficient pollination occurs.

Red clover must be pollinated by insects for producing seed. Bumblebees are usually depended upon for pollinating but they usually are inadequate in number. This is probably the number one reason for low seed yields in Missouri. Honeybees can and should be used to increase pollination of red clover. The notion that honeybees do not pollinate red clover because it has deeper florets than other legumes is false. Pollination of the flower occurs on each visit of the bee regardless of whether it gets nectar.

Provide at least one hive of honeybees per two acres of red clover. Don't place them in the field too early or they will tend to orient to other clover. Place colonies in or near the field as soon as 5 to 10 percent of the second crop is in bloom. In Missouri, this is usually sometime in July. For highest seed yields, windrow red clover for drying. Although it is occasionally harvested in the standing condition, yields obtained in this method are usually low.

Red clover should be cut for seed when the interior of the heads is brown, the stems show a yellowish tinge and the seeds show colour ranging from yellow through purple. Cutting the seed crop sooner than this results in light seed of poor viability and germination. If cutting is delayed much beyond this point, heavy seed losses occur as the dead-ripe heads are broken and lost by weather and harvesting operations.

Windrow the crop immediately after mowing. If it is allowed to dry in the swath, rake it in the morning while it is damp and tough. Seed is sometimes harvested from a new seeding of red clover made during the winter but it is usually not a desirable practice because yields are low.

Red clover quality

If red clover is harvested for hay when no more than one-half of the plants are blooming, the feeding value approaches that of alfalfa. The protein content of red clover can be nearly as high as that of alfalfa. Red clover cut for hay at the 50 percent bloom stage usually exceeds 14 or 15 percent protein. Many times red clover is not harvested soon enough; it is under these conditions that unfavourable quality results.

Don't wait until it is in full bloom. Also, when harvested too late, new growth from the crown has started to elongate. When these young shoots are removed, a sharp reduction in the second growth often occurs. Livestock sometimes slobber excessively when feeding on second-cutting red clover hay. This condition, caused by a fungus, is rather rare and usually does not harm animals. A good way to avoid this is to pasture the second growth.

Red clover insects

Insects are extremely troublesome to red clover in Missouri. They reduce yields and quality, interfere with seed production and, in some cases, destroy the stand. Unfortunately there is no practical, efficient overall control for all red clover insects. In particular, little can be done to control insects in seed fields. Because bumblebees and honeybees are largely responsible for seed pollination, chemical treatment to control insects would also destroy the pollinators. There is opportunity for economic control of some insects such as army worms and grasshoppers Both of these are capable of completely destroying red clover stands, and when populations of these insects begin to build, extreme damage may be imminent. It is then best to contact your county MU Extension center to determine the latest control methods.

Early recognition of insect damage and the insect causing it is the key to effective control. The pests are generally grouped according to the location and type of damage they inflict on the clover plant. These groups are root infesting species, stem and leaf feeders and flower head- and seed-eaters.

The root infesting species include:

— *Clover root curculio.* A short-snouted, dark gray to brown weevil or a grayish white, legless brown-headed grub about 1/6-inch long. Plants wilt and die during dry weather from feeding along the roots and girdling near the crown. Injury is most severe in late spring and early fall.

— *Clover root borer.* A very small black or dark brown, cylindrical beetle or very small, legless, curved brown-headed grub about 1/10-inch long. They feed on the root surface and tunnel through the roots, causing clover to turn brown, wilt and die. Injury is most common in old stands of clover.

— *Grape colaspis.* Clover roots are fed upon by 1/8-inch long white-curved bodied, brown-headed grub. The adult, a light-brown chunky beetle 1/10-inch long, feeds on clover leaves.

The stem and leaf feeding species include:

— *Grasshoppers.* Especially damaging during times of moisture stress.

— *Clover leaf weevil.* Small, green, legless larvae with a yellow-white stripe edged with red down the center of back. It feeds at night on the leaves. It may be mistaken for alfalfa weevil larvae when small, but grows to 1/2-inch long. The larvae hide at the base of the plants during the day. The adult has black flecks on its back and is 1/2-1/3-inch long. Both larvae and adults feed on clover leaves, leaving smooth-edged notches along the margins. Damaging populations occur more frequently in late cool, dry springs.

— *Lesser clover leaf weevil.* A small, pale brown or green, legless grub, up to 1/4-inch long, feeds in the heads and lateral buds. Buds die and plants are stunted. The adult weevil is green or bluish-green with a black head and snout approximately 1/8-inch long. The female lays eggs in small slits cut in the stem, in the bud at the axil of the leaf or in the the terminal bud of the plant.

— *Alfalfa weevil.* Damage usually occurs when adults migrate from nearby alfalfa fields that have just been harvested. Usually the alfalfa weevil is not a serious red clover pest.

— *Green cloverworm.* Light green, 1-1/4 inches long, with a narrow white stripe and a second faint white line on each side. The adults are dark-brown, black spotted or mottled moths with a wing span of about 1-1/2 inches.

— *Armyworms.* Dark green worms with longitudinal light stripes down their sides and back. Caterpillars strip foliage from clover and eat tender shoots at night but hide beneath debris during the day.

— *Cutworms.* Plump cylindrical worms up to 1-1/2 to 2 inches long, of varying shades and markings. They eat leaves and cut plants off at the base.

— *Clover stem borers.* Yellowish, smooth-sided cylindrical worms about 1/2-inch long with two curved hooks at the end of the body. The larvae tunnel inside the core of clover stems (usually sweet clover), eating out the pith. Stems swell, crack open and sometimes break off.

— *Aphids.* Small, pale green, soft-bodied plant lice. Infected plants become coated with a sticky fluid from the aphids.

— *Leafhoppers.* Elongated, wedge-shaped bugs, mostly less than 1/4-inch long and variously coloured. Leaves become mottled and white in appearance. They may become dwarfed and yellowish or reddish in colour.

— *Lygus bugs.* Greenish to yellowish-brown, these are flat bugs about 1/2-inch long. They cause blasted buds, flower drop and shriveled seeds.

— *Meadow spittlebugs.* Concealed in frothy masses on clover stems, they suck juices from the plant, causing it to wilt and die.

The flower head and seed eating insects include:

— *Clover head caterpillar.* A small hairy caterpillar about 1/2-inch long that feeds on developing seeds and destroys many florets at the base.

— *Clover seed chalcid.* Very small, fat, white, legless, maggot-like larvae that feeds inside the seeds. Infested seeds are often dull-colored and may later break and crack.

— *Clover seed midge.* Very small, pinkish, legless maggot that feeds on the outside of green seeds causing them to shrivel and dry up. Heads fail to develop evenly, only a part of the pink florets open and the rest of the head stays green.

Use of Red Clover in the Modern Period

Red clover (Trifolium pratense L.) is one of the leading forage legumes in the U.S., Canada, and northern and eastern Europe. Red clover has maintained a prominent roll in animal agriculture in the more temperate regions of the world due, in part to high seedling vigor, ease of establishment, rapid growth, excellent soil improvement characteristics, high forage quality and reasonably high yield. Records would indicate that it was first cultivated in Europe in the third and fourth centuries, recorded as being in Italy by 1550, in England by 1645, and in the United States by 1663. The early settlers brought red clover to the US as it was an excellent feed source for their animals.

As settlement moved throughout the nation red clover moved with it. Being adapted to temperate climates, red clover was and is used most prominently in the northeastern and northcentral regions of the US, and southeastern Canada for hay, silage, grazing and soil improvement. Use of red clover declined substantially beginning in the 1950s as alfalfa (Medicago sativa L.) became more prominent forage for hay and silage. Nationwide, red clover and grass mixtures declined from approximately 15 million to 12 million acres between 1950 and 1970, and based on seed disappearance they probably declined to about 11 million acres by 1987. However, with the increase amount of grazing over the past decade the use of red clover has increased.

Improvement of Red Clover

Red clover has long been considered to be a short-lived perennial. Environmental stress and root rot diseases contribute to the lack of persistence of red clover. Of the root rot causing pathogens of red clover, organisms of the

Fusarium species have been isolated most consistently from diseased roots. The foliar disease, northern anthracnose, caused by Aureobasidum caulivora (Kirchn.) W.B. Cooke, is endemic and causes severe forage loss during the first growth of the season. Therefore, resistance to these pathogens should lead to improved persistence and yield. Selection for plant longevity in older stands of red clover and the incorporation of genes for resistance to root rot pathogens are the most practical methods of improving persistence of red clover.

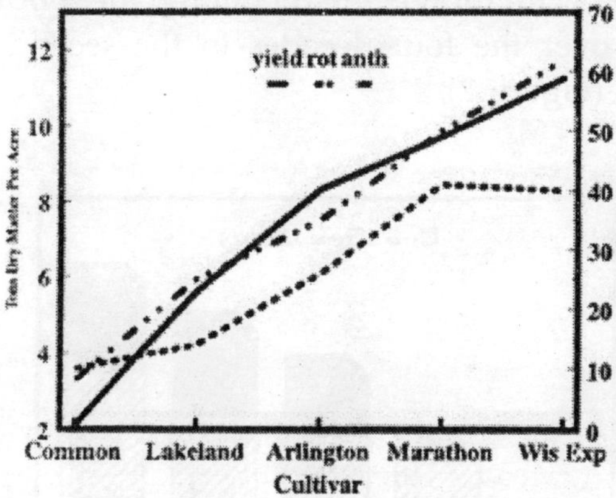

Figure 1. Progressive improvement in the performance of red clover resulting from four decades of breeding for persistence and disease resistance

Earlier, USDA—Agricultural Research Service—University of Wisconsin project has made in red clover through selection for persistence and disease resistance since the 1950s. Both forage yield and resistance to root rot and northern anthracnose has been improved (Figure 1). Resistance to the two diseases, root rot and northern anthracnose, has steadily increased with the development of new germplasm. Two to three cycles of selection for resistance to northern anthracnose was applied during each cycle of selection for persistence. On the other hand, only one cycle of selection for root rot occurred after the third cycle of selection for persistence.

Earlier improvement in root rot resistance was the result of natural selection for resistance with the selection of healthy, persistent plants. Total forage yield increased over the cycles of selection due, in part, to selection for disease resistance and to increased persistence. There was no difference in forage yield among the populations developed over the four decades in the second year of production (Figure 2).

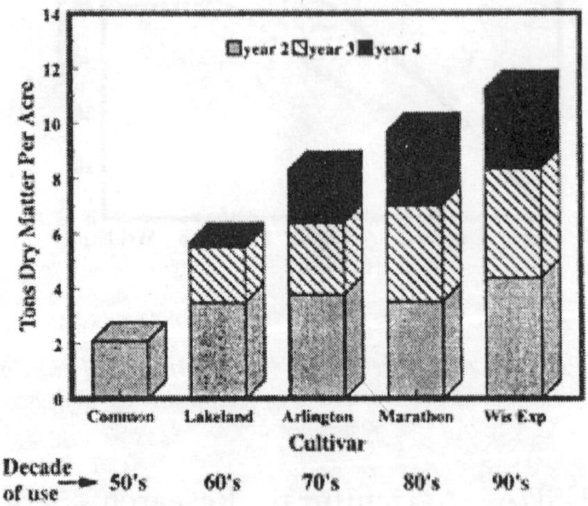

Figure 2. Forage production of red clover resulting from five decades of breeding for persistence and disease resistance

The differences among the populations occurs in the third and fourth year of production with Marathon and Wis Exp producing excellent forage yields in these latter years

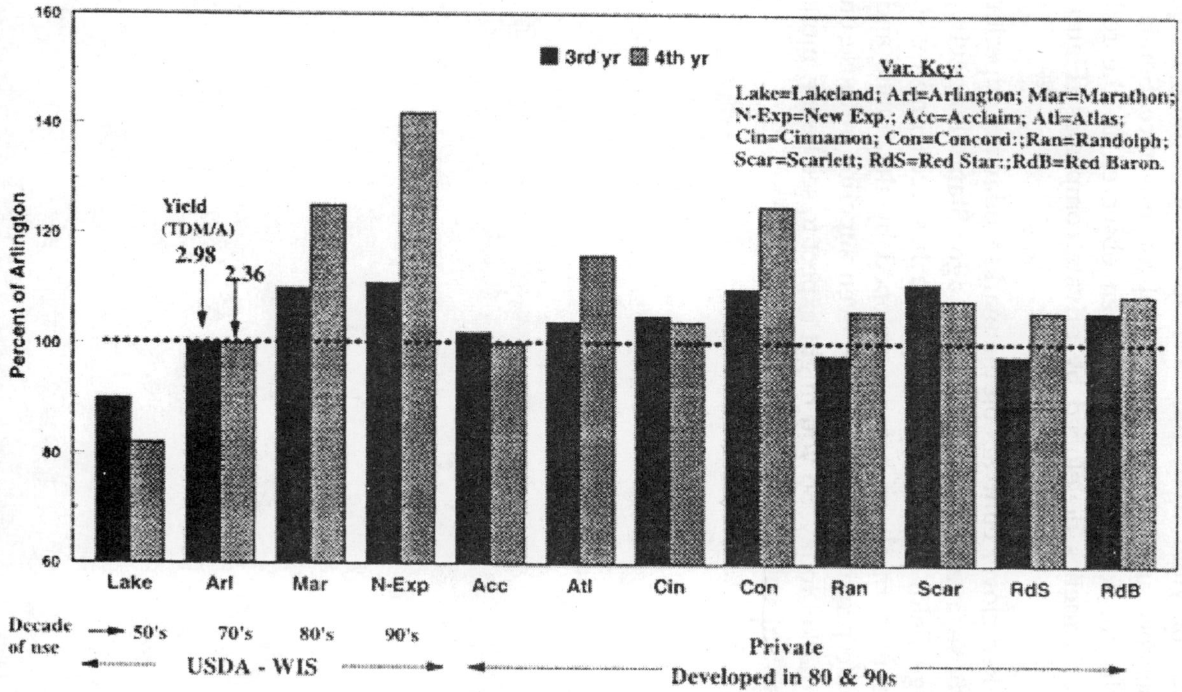

Figure 3. Forage yield of red clover varieties grown in wisconsin in 3rd 4th year of production relative to arlington

of production. Selection in exiting 3- and 4-year old stands of yield trials has been effective in improving persistence and to some extent root rot resistance. Similar results for forage yield have been observed in the red clover varieties developed by private companies (Figure 3).

Red clover cultivars on the market today are superior to those available two decades ago. Stand longevity, forage field, and disease resistance has been improved. Current germplasm being developed by the public and private programs appear to be even superior to those on the market today. So you might expect to see better, more persistent cultivars in the future.

13

Grass Clover

Grass-legume pastures are the basis of the dairy industries. The grasses used include annual and perennial ryegrass, tall fescue and cocksfoot. Red and white clovers form the legume component. While pure grass pastures generally have higher dry matter production per hectare than pure clover pastures and provide for a more even fodder flow, clover pastures provide a higher quality fodder than grasses. With good management a high proportion of clover can be maintained in grass pasture without compromising the total dry matter production of the pasture. With grass-clover pastures not only is forage quality increased but, through their process of fixing atmospheric nitrogen and the recycling of fixed nitrogen, clovers reduce the requirements for inputs of inorganic nitrogen.

The production potential of a perennial grass-clover pasture depends mainly on the stability of the grass-clover ratio. For optimum quantity and quality of herbage a clover content of 30-50% is needed. As the contribution of clovers to total dry matter production of the pasture increases above 50%, so the dry matter production per hectare is significantly reduced. If, however, clover's contribution is below 30% then there is a loss in quality of dry matter availability to livestock. For example, cows grazing pastures which have a 25% or 50% clover component produce 22% and 33% more milk respectively than pastures which have a 0% clover component at the same dry matter availability per hectare.

Due to the different growth forms and growth requirements of perennial temperate grass and of clovers, and differences in palatability of the plants, most farmers find it extremely difficult to maintain a sufficient percentage of clover in their pastures. Grasses tend to be vigorous growers and, given the opportunity, will out-compete the clovers. The management of grass-clover pastures therefore puts greater emphasis onto ensuring that the requirements for optimal growth of clovers are maintained.

In Denmark grazed grass/clover pasture is considered to be a cheap source of animal feed due to the absence of harvesting costs and the low N-fertilisation requirement. With continuous grazing of grass/clover pasture the sward height, the area rejected for grazing and milk production per ha and per cow are good indicators of the management level. If, for example, the grazing intensity is low, then the herbage height, rejected areas, intake and milk production per cow will be high, but the milk-production per ha will be low, because much herbage will senescence before it can be eaten be cows. The management system takes into account the general variation in herbage-production within and between the years in accordance with plant available water and fertility on the given soil-type.

Other factors in focus are clover content; the level of N-fertilisation; irrigation and the area used for silage making. The main goal of the system was to maintain an optimal balance between milk production and utilisation of herbage fields under practical conditions. The strategy was to fulfil the milk-quota with a minimum of cost. The cost-minimisation include a maximal intake of grazed grass, a minimum of silage and a low N-fertilisation level compensated with N-fixation by clover.

Tools for Management

On commercial farms, the aim of optimising the utilisation of grassland must be incorporated into a complex fanning system that imposes additional objectives and constraints. The grassland area is only a part of the roughage production on a dairy farm in Denmark; typically 3-5 crops arc grown in a rotation system. The grassland management is dependent on and interacts with these other crops. The strategy is often to use the grassland solely for grasing; winter feed is supplied by whole crop cereal silage and fodder beet.

In Denmark, the grass/clover is normally undersown in cereals, which often is used for whole crop-silage for winter feeding. Normally the aftermath after cereals will have a high clover content, when low amounts of N are used. The grass/clover sward will normally be utilised for grazing in the autumn and the following two years. Then the sward will be ploughed for cereals or fodder/ sugar beet. If the clover disappears, the farmer can either use more N-fertiliser or exchange the field with a 2-year old grass/clover-sward with a good clover stand.

The silage cost for making grass/clover silage is higher than the cost for making silage from whole crop-cereals, leading to a strategy for maximal grazed amount of herbage, and a minimum of grass/clover silage. The area for whole crop silage is adjusted in accordance to winter feed demand. The pasture production based on grass/clover with moderate N-fertilisation in a crop rotation system, and the costs of fencing favour a system with continuous stocking. Control of the grassland utilisation under these conditions aims to maximise utilisation of the daily herbage growth by grazing to fulfil the given milk quota, at lowest cost of fertiliser and supplemental feeding.

The close connection between milk yield per cow and herbage production per ha makes it necessary to incorporate milk production and herbage utilisation in one system. Uncontrollable factors such as climate make it impossible to incorporate all the details in the tactical plans. For this reason the system starts off with a grassland budget and summer feed plan, which include the tactical goals for production of milk and silage and the expected growth rate of grass/clover. The tactical goals are the guidelines for the control at the operative level. This relationship is based on "the iterative approach to the farm planning process" described by Sørensen et al. It is therefore essential that the system is self-adjusting through a constant check on the grass and milk production.

The control on an operative level is based on daily registration of the grass/clover condition, sward height and the areas with rejected grass. The registrations are compared with the goals, made during the planning process of the grassland utilisation. In the case of disagreements between the registered conditions and the goals, adjustments are made in relation to the alternatives set up in the plan.

Here we present a management system for tactical/operational planning and control on commercial farms. The tool is developed to include optimum utilisation of a given grass/clover area through a growing season, continuously stocked. The system is developed to be used by advisers as well as farmers. On the basis of the theory of the iterative approach to the farm planning process a management system on mainly pure grass has been developed and tested from 1987 to 1989.

Tools for Planning and Adjustment

Tactical strategy for feeding and grassland utilisation

On the tactical level a plan for summer feeding and the overall grassland utilisation was made during the winter.

Tactical goals were specified for following periods:

Spring:	May (until 1st grass/clover silage cut).
Summer:	June-July (until harvest of 2nd cut, grass/ clover silage or whole crop cereals).
Late Summer:	August-September (until start of supplemental feeding during night-time).
Autumn:	September-October (until start of winter feeding).
Winter:	November to April

Since the end of the first two periods correspond to the normal time for silage cuts, it was therefore possible to adjust the area for grazing at the beginning of these times. For each period a grassland utilisation budget was calculated based on a typical net production rate of grass/clover and animal requirements for herbage. This estimate is dependent on soil type, irrigation, grass/ clover condition, N-fertilisation and season.

During the planning process, the amount and time of season for N-fertilisation, and use of either silage or grazing, were used to manipulate the growth rate in favour of a high amount of the pasture production for grazing. Feed rations for the dairy cows were formulated in relation to herbage intake and target production level.

The herbage supply from both the pasture area and the area with grain undersown with grass/clover is estimated. The grain area can be used either for cereals or for whole crop silage for winter feeding, and the aftermath can be grazed. In periods with both silage and grazing, grazing was normally recommended in the fields with the highest clover content and highest tiller/stolon numbers. Herbage planned for silage was normally supplied with the highest level of N and this, together with undisturbed growth for cutting, is the reason for a higher expected growth potential than from the grazed sward.

Pasture intake per cow daily was calculated in relation to hours on pasture, clover content in pasture, season and total energy level. During the planning process, the amount of supplementary feed was manipulated in order to reach a high fresh herbage intake and to adjust milk production in relation to milk quota. The planning process is then iterative, where the goal is for each of the four periods to balance the herbage production with the herd intake, in order to fulfil the overall goals in relation to milk quotas and minimisation of the costs of concentrates and N-fertilisation. This is achieved by trying to manipulate some of the factors mentioned earlier. The system does not include any specific rules to optimise the plan. Often a lot of practical factors, such as field size, maximum number of cows, silage facilities etc., have a significant impact on the optimal plan, under practical conditions.

The planning, therefore, has to be prepared by an adviser, with a good theoretical background, and the farmer. A very important part of the planning process is to set up alternative plans, if the average conditions should not be reached. In the example the grassland area is sufficient for maximum herbage intake. As the growth rate declines during the season, the grazed area is increased. In the autumn, night feeding with silage and fodder beets compensate for the declining herbage growth rate. During the season the expected herbage intake per cow is 1600 SFU, or around 50 % of the total energy demand. The strong focus on a high amount of fresh herbage is expected to result in only 13 percent of the herbage production for silage, while the rest is utilised by grazing.

Control and Adjustment

Control of the tactical plan is done in two ways. On a daily basis the balance between herbage production and herbage intake has to be controlled in order to get a satisfactory utilisation of the pasture area. On a periodic basis, the tactical goals for herbage production, amount of supplemental feeding, and area for grazing can be controlled in order to get information to be used in the tactical planning for the next season.

The daily control has to be done by daily inspection of the sward. Sward height is measured by the farmer, and the amount of rejected area is calculated. If the actual conditions differ from the plan, the expected balance between herbage production and herd demand will not be reached. In principle imbalance can be a result of three factors: supplemental feeding, utilised herbage area, or herbage production per day. If one or more of these factors are in disagreement with the plan, it will result in an incorrect grassland utilisation.

In principle the daily adjustment, on the basis of sward condition, can be made in two ways:

1. Adjusting the grazed area, or
2. Adjusting the feeding level of supplemental feeds.

In Table 1 the recommended principles for grassland utilisation. The size of adjustment and nature of the adjustment (adjusting area or supplemental feeding) are decided by the farmer. Optimum utilisation of the pasture is assumed to be related to sward height and/or amount of rejected area. Adjustments are necessary if the pasture deviates from either target. Working with continuously stocked pasture, it is important to focus on both factors.

Table 1. Principles for grassland utilisation at continuous stacking with dairy cows.

Recommendation for sward height in grazed area and area with rejected (mainly fouled) herbage.

Weeks after start of grazing	Planned sward height min. - max.	Planned area with rejected min. - max. herbage
0- 8	5 - 7 cm[1])	0-5%
8- 16	4 - 6 cm[1])	10 - 15%
16-24	4 - 5 cm	15 - 25%

1) In older swards with high tiller/stolon density down to 3-4 cm sward height can be accepted before adjustment is needed.

Herbage shortage. Sward height or amount of rejected area less than planned. → Use feed plan "shortage*" or expand grazed area.

Herbage surplus. Herbage height or amount of rejected area more than planned.

- Rejected area >5% above planned → Use feed plan "surplus*" or close up grazed area.
- Rejected area >10% above planned → Top the paddock or graze by young stock or silage after 3-6 weeks rest.

Under some circumstances (e.g. the start of the grazing period), it is easy to detect deviations from the target sward height while it is more difficult to observe rejected areas. Later in the season, lax grazing will often first be registered by a deviation from the target amount of rejected area. The periodic control is done by monthly registration of number of cows, milk production, amount of supplemental feeding, utilised areas for grazing, and silage production. From these factors net herbage production per ha per day can be calculated indirectly. This control is used to adjust the planned amount of silage from whole crops and used to adjust the expected herbage production, when making plans for the next season.

14

Aliske Clover

Alsike Clover introduced perennial plant is ½-2' tall, branching occasionally. The stems are round and hairless, or nearly so. Alternate compound leaves occur along these stems. They are trifoliate, and have long petioles. Where the petiole joins the stem, there are a pair of membranous stipules that are broadly lanceolate. The individual leaflets are about ½" long, and either oblong or obovate in shape. They are neither sharply pointed nor indented at their tips, while their margins are finely serrate.

Unlike some *Trifolium* spp., there are no chevrons (white markings) on the upper surface of these leaflets. The foliage is hairless. From the stipules of the upper stems, there are flowering stalks that terminate in well-rounded flowerheads about ½-¾" across. Each flower is about ¼" long and very slender, consisting of a tubular corolla and 5 lobe-like petals. Like other pea-like flowers, there is a standard, two side petals, and a keel, but they are tubular and narrow. The calyx of each flower is white with long slender teeth that are green. It is much shorter than the corolla. The flowers are usually some shade of whitish pink. The blooming period occurs intermittently during the summer for 1-3 months. The root system consists of a taproot and secondary roots that form nodes with nitrogen-fixing bacteria.

Features of Aliske Clover

Cultivation

The preference is full or partial sun, and moist to mesic soil that is slightly acidic. This plant flourishes in a loam or clay-loam soil.

Range & Habitat

Alsike Clover is a common plant that occurs in every county of Illinois. However, it appears to be less common than either *Trifolium pratense* (Red Clover) and *Trifolium repens* (White Clover). Alsike Clover was originally introduced from Europe, probably for agricultural purposes. Typical habitats includes moist meadows near woodlands, pastures, abandoned fields, and roadsides. It does not tolerate regular lawn-mowing as well as White Clover.

Faunal Associations

Long-tongued bees visit the flowers for pollen or nectar, including honeybees. Many kinds of insects feed on the foliage, including the caterpillars of the butterflies *Colias philodice* (Clouded Sulfur) and *Colias eurytheme* (Orange Sulfur). The caterpillars of many moths feed on the foliage and other parts of Trifolium spp. Because the foliage contains oxalic acid, it is mildly toxic to some mammalian herbivores if it is eaten in sufficiently large amounts. Nonetheless, livestock, rabbits, and groundhogs readily consume it, like other clovers. The foliage, flowerheads, and seedpods are eaten by upland gamebirds, including the Greater Prairie Chicken, Wild

Turkey, and Bobwhite. The Horned Lark and Thirteen-Lined Ground Squirrel also eat the seedpods.

Alsike clover has been an important summer legume in north central and Northeastern United States; however, it is being replaced by alfalfa. A small amount of seed is produced in the Pacific Northwest. Most of the seed that we use comes from Canada. In 1969, we imported 3,715,000 pounds, most of which was probably used in mixed pasture planting.

Alsike clover is a perennial but usually grows as a biennial, and in some situations it behaves as an annual. It is intermediate in size between white and red clover. Many smooth stems, bearing smooth trifoliate leaves, arise from its crown. The noncreeping stems may grow to a height of 5 feet but usually reach about 2 feet. They bear flower heads along their entire length, the youngest always toward the top. It is a good hay, pasture, and green manure crop, and like other legumes, it improves the soil and contributes to reduced soil erosion.

Inflorescence

The flower head of alsike clover is made up of many florets and is similar to the more common white clover, although there may be four times as many heads per square yard as are normally found on white clover. The florets on some heads are pink, on others they are white, and on some they are both pink and white. This variation in flower color led to an earlier belief that alsike was a cross between red and white clover, hence the scientific name hybridum.

A floret will produce two to three seeds. Flower heads with 100 seeds indicate a good seed crop. The flowers are quite attractive to bees, especially honey bees, for the nectar and pollen. Pellett stated that some beekeepers estimated that alsike clover might produce 500 pounds of honey per acre in good seasons. This seems abnormally high but indicates the importance beekeepers attach to this crop as a source of honey. Holmes noted that boron favorably influenced nectar secretion in alsike clover, but apparently this information has not been used to increase seed production.

Seed Production

Alsike clover is intermediate between red and white clover. Plants are semi-erect growing tall enough to be cut for hay. Leaflets are not marked with a white V te flowers vary from white to almost completely rose colored. Seeds vary from yellowish to dark green in color. Alsike clover has been grown in Europe for centuries an it is an important legume in the timothy-clover area of the US and Canada. It is used as a compliment to Red clover in meadow mixtures. It tolerates more flooding, acidity, and alkalinity than other clovers.

Alsike clover usually acts as a short-lived perennial, but in some regions (e.g. - northern Alberta), it persists for many years. It seems likely that its short lifespan under other conditions is due to root diseases. It grows best in cool climates, and on low-lying wet or peaty soils that may flood for short periods. Alsike tolerates soils too acid or too saline for red clover. Most varieties are winterhardy, though the tetraploid type is less so than the common diploid type.

The seeds are very small and should not be seeded more than one half inch (1.25 cm) deep. Pack drilled seedings with a wheel press and broadcast seedings with a surface packer. Seed one to two lbs/acre (1.1 to 2.2 kg/ha) of inoculated seed in a row spacing of 12 in (30 cm). A companion crop is not recommended, but if one is used, seed it at about one-half the usual rate in 12 in (30 cm) rows and crosswise to the clover. Spring seeding is recommended, but stands can be established as late as mid-July if soil moisture is adequate and the weather is not too hot and dry.

Fertilizers

Fertilizers are not usually needed, but a soil test is recommended and any required fertilizer should be worked in before seeding.

Pollinators

Alsike clover is a favourite pasture for honey bees and they do an excellent job of pollination. At least one strong colony per acre is recommended for large fields and more than one colony per acre for small fields.

Insect Control

Severe insect damage is unusual. If an insecticide is needed, apply it when the bees are not in the field or remove the hives during the spraying operation. Ideally, the crop should be sprayed before the plants are in flower. Remove crop residue after the harvest to aid in the control of insects and diseases.

Harvesting

Alsike clover matures earlier than red clover, and usually can be harvested late in July or early August. The seeds shatter readily, so handle the crop as little as possible. Straight combining is not advisable, since seed can easily shatter before the stems and leaves are dry. It is usual to harvest with a swather or a mower with a windrowing attachment, and to cut when the plants are wet with dew or rain. Harvest when a majority of the seeds are ripe or turning purplish. Pick up and combine the swaths before they are completely dry, to avoid losses due to seed shattering. Combining usually occurs after five to seven days of good drying weather.

Seed Cleaning

Top screen 1/19; bottom screen 6 x 32.

Stand Life

Red clover will normally be productive for seed production for a maximum of 2 seed harvest years.

Adaptation

This cool-season crop is adapted to low-lying, moist areas. It will tolerate soils that are completely waterlogged and withstands spring flooding up to six weeks. It is also well suited to acidic, organic soils. Alsike clover tolerates more alkalinity than most other clovers. It is easily established where there is minimal land preparation, but must be seeded shallowly. Of our legumes, it is the most tolerant of cold and frost heaving. Damage from insects and diseases and depletion of root carbo-hydrates are uncommon, so that it has excellent winterhardiness.

Use for Hay

In the moister areas of Montana, alsike clover yields well, and generally will thrive where other legumes fail. Like red clover, it is high in moisture content, and therefore, is difficult to cure in the field; however, it retains its green color somewhat better. Alsike clover is seldom grown alone, but it produces good yields in mixtures with grasses such as timothy, and since the grass holds the clover more upright, harvesting is easier. Normally, only one cutting can be harvested for hay each season.

Use for Pasture

Regrowth after hay cutting is quite good and is similar to that of single-cut red clover. Alsike clover is very palatable to cattle. Proper fall use will not deplete root carbohydrates or affect winterhardiness. Alsike clover is satisfactory in a pasture mixture, although its short life limits its usefulness to the first few years of production. It is somewhat difficult to control the proportion of alsike clover in a mixture, since this legume tends to dominate the stand for the first one or two years, and then it decreases rapidly. The bloat hazard is similar to that of red clover or alfalfa.

Limitations

Alsike clover is intolerant of drought and high temperatures. It will not survive on land that floods in spring, but dries up in the heat of summer. Salinity tolerance is also low. Shade tolerance is poor which makes it less useful for mixtures with tall growing grasses such as reed canarygrass.

15

Crimson Clover

Crimson clover (*Trifolium incarnatum*), also known as Italian clover, is a species of clover in the family Fabaceae, native to most of Europe. The species name incarnatum means "blood red". This upright annual herb grows to 20-50 cm tall, unbranched or branched only at the base. The leaves are trifoliate with a long petiole, each leaflet hairy, 8-16 mm across, with a truncated or bilobed apex. The flowers are produced throughout the spring and summer, rich red or crimson, congested on an elongated spike inflorescence 3-5 cm tall and 1.5 cm broad; the individual flowers are up to 10-13 mm long and have five petals. The banner of each flower does not sit upright, but folds forward.

Development of Crimson Clover Cultivation

Crimson clover growing in Texas. Crimson clover is widely grown as a protein-rich forage crop for cattle and other livestock. It can typically be found in forest margins, fields and roadsides. It has been introduced into the United States, originally as forage for cattle. It is often used for roadside erosion control, as well as beautification, even though it tends to eliminate all other desirable spring and early-summer species of native vegetation in the area which it is planted.

Duke listed the following as reseeding varieties of crimson clover: 'Dixie,' 'Auburn,' 'Chief,' and 'Talladega,' and 'Tibbee;' 'Frontier' is a soft-seeded variety. According to Knight, five named reseeding cultivars have been used: 'Dixie,' 'Auburn,' 'Autauga,' 'Chief,' and 'Talladega;' the first three are mature about a week earlier than the last two. Cv 'Flame' has a higher proportion of hard seed, matures earlier, and produced more biomass than 'Dixie'.

Seed Description

Crimson Clover has about 330,000 seeds/kg. Hardseededness is a character that reduces the chance of premature germination and ensures self-reseeding. Among varieties of crimson clover, hardseededness can vary from 30-75%. Five named reseeding cultivars have been used: 'Dixie,' 'Auburn,' 'Autauga,' 'Chief,' and 'Talladega.' The first three are mature about a week earlier than the last two. 'Flame' has a higher proportion of hard seed than 'Dixie'.

Mature Plant

Crimson clover is similar in appearance to rose clover but is a taller, more erect plant. The showy crimson flower also distinguishes the crimson clover plant. The light-green foliage is covered with soft hairs; the leaves are usually unmarked but sometimes have a few dark-red spots; the plant branches less than does rose clover, and can grow to 1.5 ft. in height. Crimson clover flowers are self fertile but not self pollinating. Flowers produce much nectar and are heavily visited by various types of bees.

Temperature

Crimson clover cannot tolerate extreme heat or cold and grows best during the cool, humid weather that typifies the winter in mild climates. It tolerates mean annual temperatures of from 5.9-21.3 degrees C, with the mean of 30 cases being 12.9. Seedlings require considerable heat in the early stages of growth. Crimson clover should be established 6 weeks before the average date of the first frost.

Williams found that the forces exerted by crimson clover and alfalfa seedlings, which can be important in overcoming soil crusts, compaction, and excessively-deep seeding, were highly dependent on temperature, with the maximum force attained for crimson clover near 20°C, and for alfalfa between 25 and 30°C. These patterns suggest that plantings could be timed to achieve best emergence of seedlings. In establishing crimson clover, moisture tended to be limiting through November 1, with temperature the critical consideration thereafter, until mid-February.

Geographic Range

Crimson clover is native to Atlantic and southern Europe, Caucasus and Transcaucasus. Knight and Hoveland maintain that it is native to southern Europe. It is assigned to the Mediterranean and Eurosiberian Centers of Diversity, is grown from the subtropical through cool temperate regions, and ranges from the Boreal Wet through Subtropical Moist Forest Life Zones. In Europe, it was cultivated for forage and green manure during the 18th century. Crimson clover appears well-adapted to the predominant climatic conditions, and it tolerates a wide range of climatic and soil conditions.

Water

Crimson clover tolerates from 3.1-16.3 dm (12.2-64.2 inches) of annual precipitation, with the mean of 30 case being 9.2 Duke, and grows well in regions with 8.89 dm (35 inches) or more annual rainfall. It cannot endure much drought and likewise does not do well on poorly-drained soils. Moist soil is essential for germination and establishment. In California, crimson clover will reseed only under excellent moisture conditions, i.e., with moisture at least throughout April.

In establishing crimson clover, moisture tended to be limiting through November 1, with temperature the critical consideration thereafter, until mid-February. Zachariassen and Power, found that crimson clover showed a consistently-higher water use efficiency (g of dry matter produced per liter of water evapotranspired) than hairy vetch at 10, 20, and 30C. Sweet clover showed intermediate values.

Nutrients

Crimson clover is adapted to soils of low fertility, and has an intermediate lime requirement; on infertile sites, phosphate and potash fertilisers and manure may increase production. A fall application of 200-400 kg/ha of 0:20:20 N:P:K fertiliser may suffice for two years on infertile sites.

Soil pH

Crimson clover is tolerant of alkaline soils and of a pH range of 4.8-8.2, with the mean of 26 cases being 6.5. Slayback (pers. comm.) suggested a pH range of 5.5 to 7.0. Crimson clover is intolerant of poorly-drained calcareous soils, and muck or extremely acid soils do not support good growth. Crimson clover tolerates a wide range of soil conditions but does not do well on calcareous soils or with poor drainage. Duke wrote that crimson clover or varieties thereof are noted for tolerance of high pH, heavy soil, and sand, thriving on sand or clay soils, so long as drainage is good.

McLeod stated that the species will grow on almost any soil, provided it is well-drained, but said that best growth occurs on loam soils with a good humus content. Soils that are cold or waterlogged during the winter are not suitable nor are muck or extremely acid soils. Other recommendations on suitable soil types include: sandy loam to clay loam (Slayback, pers. comm.), clay or sandy soil; and Ultisols.

Life Cycle

Crimson clover a stout, soft-pubescent winter annual. Seed should be sown from mid July until November, with early seeding more important in northern sectors; spring plantings usually result in stunting, poor flowering, and low seed yields. Crimson clover grows in the winter but puts on most of its growth in the spring, maturing in May. Flowering is induced when daylength exceeds 12 hours. The flowers are self fertile but not self pollinating, produce much nectar, and are heavily visited by various types of bees.

Seeding Rate

Seeding rates have been variously suggested as 9 lbs/acre; 12-15 kg/ha (10.7-13.4 lbs/acre); 15 to 20 lbs/acre; 15-30 lbs/acre; and 40-50 lbs/acre for unhulled seed. Crimson clover can be grown in conjunction with rye, vetches, annual ryegrass, and various cereals. When sown in combination with companion crops, crimson clover is usually sown at 2/3 the normal rate, and the other crop at 1/3 the monocultural rate. Developmental rates of annual ryegrass and tall fescue are similar to that of crimson clover.

Seeding Method

Duke suggested a firm seedbed and broadcasting or drilling; drilling produces a better stand than broadcasting. If seed are drilled, spacing between plants should be about 2.5 cm (1 inch). Crimson clover can be grown in combination with cereal grains, annual ryegrass, and other winter annual legumes. It can also be established in existing stands of bermudagrass, dallisgrass, johnsongrass, bahiagrass, or sericea lespedeza. Before overseeding into existing grass sod, grass must be mowed closely to allow seedlings to establish. Light disking can also reduce competition by grasses.

Seeding Dates

Seeding can occur from July until November, with early seeding being important in northern sectors. The aim of early sowing is to ensure that plants are well established before freezing weather occurs; frost heaving is especially damaging to young plants. In fact, crimson clover should be established 6 weeks before the average date of the first frost. However, July sowing can result in stand failure due to virus. In Mississippi, August 15 planting led to the highest yields. In establishing crimson clover, moisture tended to be limiting through November 1, with temperature the critical consideration thereafter, until mid-February.

For California, fall seeding is suggested. Slayback (pers. comm.) specified that September-October seeding dates are best and that spring sowing for summer production necessitates irrigation. Spring seeding often results in stunting, poor flowering, and reduced seed yield. Despite its natural winter-annual growth habit, in the Far North, crimson clover can be spring sown and grown as a summer crop, as is done in rotation with potatoes in Maine.

Flowering is induced when daylength exceeds 12 hours. In California, crimson clover flowers from May through August, according to Munz. Cvv 'Dixie' and 'Frontier' are among the earliest to head out and set seed, but cv 'Flame' flowers still earlier (about 2 weeks). Crimson clover was considered early maturing by Finch and Sharp, and will set seed by late May.

Seed Production

Crimson clover reseeds well when allowed to mature. This has been seen in in vineyards in Del Rey (east side of the southern San Joaquin Valley, California) (Masumoto, pers. comm.). However, reseeding requires moisture at least through April, as recounted by Murphy et al. The seedheads are easily accessible for use by grazing animals, and removal of all seed-producing heads must be prevented if reseeding is to be assured.

Duke listed seed yields as commonly ranging from 340-410 kg/ha, with higher values on soils of intermediate as opposed to high fertility. Maximum seed yields are in the range of 1,000-1,200 kg/ha. According to Duke, seed can be harvested in three ways:

(1) combine standing plants;

(2) mowed and left to swath or windrowed to dry and later threshed with a combine; or

(3) mowed, swathed, or windrowed to dry, then hauled to a stationary huller or thresher. Hulls should be dark brown if seed is to be combined directly, light brown if mowed.

Pest Effects, Insects

Flowers of crimson clover are attractive to bees. In California, crimson clover flowers contain abundant minute pirate bug (*Orius tristicolor*), an important beneficial insect that preys on various agricultural pests (Bugg, pers. comm.).

In southern Georgia, crimson clover was found to sustain pea aphid (*Acyrthosiphon pisum*) and blue alfalfa aphid (*Acyrthosiphon kondoi Shinji*) which do not attack pecans. They were found along with associated predators that can disperse to pecan trees and attack pecan aphids. Crimson clover is being used by some pecan growers in efforts to enhance early-season biological control of pecan aphids.

House and Alzugaray reported that hairy vetch (*Vicia villosa Roth*) sustained higher densities of above-ground arthropods and a more taxonomically-diverse entomofauna than did crimson clover (*Trifolium incarnatum* L.) or wheat (*Triticum aestivum* L.). Soil arthropods were more diverse under no-tillage than under tillage. Pest and beneficial soil arthropods were most abundant in no-tillage corn preceded by hairy vetch. Differences that were apparent early had dissipated by midseason.

Use of Crimson Clover

Crimson clover is the most important winter annual legume in the Southeast; it is excellent for pasture, hay, or green manure and at protecting the soil during winter and spring. It can be winter grown as a green manure in rotation between milo and cotton, for example, or soybeans, small grain, and cotton.

As recounted by Murphy *et al.*, it is often used as a hay crop or for roadside plantings as well as for range seeding. It is best suited for short-rotation planting, for use in a mixture with grains, or as a rotation crop with grains, and it is sometimes cut for hay. The seedheads are easily accessible for use by grazing animals.

Crimson clover is a good green-manure crop for pecan and other orchard crops and can also be used in agroforestry, double-cropping, and no-till farming systems. In Georgia, crimson clover can be a self-reseeding cover crop in pecan, peach, and other orchards. In Alabama, it has been used to improve the soil for cotton.

Crimson Clover in Various Crops

Crimson in Corn

Crimson's higher than average shade tolerance means that seedlings survive in corn when planted at the V4 to V8 corn leaf stage. Until the canopy opens as the corn matures, there is little clover growth and, therefore, no significant competition with the growing corn plant. Clover growth following corn harvest can be rapid, and in a warm, moist year beautiful, bright red flowers may develop prior to first freeze.

Crimson in Soybeans

In soybeans, overseeding by highboy at leaf drop is recommended.

Crimson in Wheat

Unlike red clover, crimson does not do well frost-seeded into wheat or other already-established small grains. Instead, crimson does best when planted after wheat is harvested.

Crimson in Short-season Crops

Crimson clover has been used effectively to suppress

weeds when planted in the early fall, following a short season crop such as potatoes, snap beans or vegetables. Planting with a grass such as oats, which is also fast-growing, gives additional weed control in these situations.

16

Genetics of White Clover

Genetic Diversity and Conservation

Genetic diversity and conservation is crucial to meeting future human needs. Biodiversity can be preserved in natural communities and its conservation and sustainable use is likely to be a worldwide priority for generations to come. The term "biodiversity" was introduced in the middle of 1980s by W.G. Rosen. It was defined in the Convention on Biological Diversity (CBD) as follows:

"the variability among living organisms from all sources including, inter alia, terrestrial, marine and other aquatic ecosystems and the ecological complexes of which they are part; this includes diversity within species, between species and of ecosystems".

Biodiversity represents all the levels of diversity in biological organisation, including community, species and organism. It also includes differences in genes within species, difference in chromosome structure and in DNA, which determine the uniqueness of species and individuals. For living organisms, diversity is a strategy for survival and adaptation to specific environments. For humans, diversity, especially of plant species, is an essential resource for the development of sustainable food sources providing new and distinctive characters, such as resistance against disease and environmental stress.

The human population depends for half its energy directly obtained from plants on just four species wheat, maize, rice and potato. The intensive use of a limited number of improved varieties has narrowed the genetic base of important food crops leading to genetic vulnerability. Effective use of genetic diversity in agriculture is crucial to the maintenance of food supplies to feed the increasing human population in an environmentally sustainable way.

The conservation and sustainable use of plant genetic resources are considered important from public, scientific and political viewpoints. The maintenance, characterisation and use of biological diversity are also important fields of scientific research. In the early twentieth century, the Russian geneticist and plant geographer, N.I. Vavilov, recognised the importance for crop breeding of the availability of a gene pool with a broad range of genetic diversity. He directed numerous collections of crops and their wild relatives throughout the world and his work became the basis for systematic conservation and use of plant genetic resources.

Ex situ conservation, i.e. conservation of plant genetic resources out of their natural habitat, was scientifically developed in the 1960s and 1970s. The need for this led to the establishment of the International Board for Plant Genetic Resources (now International Plant Genetic Resources Institute) in 1974. In recent years, there has been an increasingly realisation of the need to halt the loss of diversity of plant genetic resources and enhance its utilisation. This idea was given emphasis by the CBD.

International cooperation on plant genetic resources has been developed, primarily based on the framework of the Food and Agriculture Organisation (FAO) of the United Nations (UN). The CBD was signed by 150 government leaders at the United Nations Conference on Environment and Development (UNCED) held in Rio de Janerio, Brazil in 1992 (Rio Summit) and since then 175 countries have ratified it. It was conceived as a practical tool for translating the principles of Agenda 21 (a comprehensive plan of action to reduce human impact on the environment which was adopted at the UNCED Rio summit) into reality.

The CBD is one of the three Multilateral Environmental Agreements to have emerged from the United Nations Conference on Environment and Development process. The main idea of CBD is not only to illustrate the need for the conservation of biological diversity but also to link conservation to its exploitation and development for human benefit. The CBD is a framework for global action on biodiversity-related issues. It has a comprehensive mandate, which is reflected in its three main objectives:

The fair and equitable sharing of the benefits arising out of the utilisation of genetic resources, including appropriate transfer of relevant technologies and application to biodiversity at three levels: genes, species and ecosystems. The CBD translates these guiding principles into binding commitments in its 15 substantive provisions. In its Article 2, *in situ* conservation is defined as "the conservation of ecosystems and natural habitats and the maintenance and recovery of viable populations of species in their natural surroundings and, in the case of domesticated or cultivated species, in the surroundings where they have developed distinctive properties".

The provisions of the CBD include measures of in situ and ex situ conservation. Scientific work in conservation has a national and international socio-political impact. Therefore, in my view, it is important to engage in this debate with knowledge to ensure that policy formulas are appropriate. The objectives of the CBD; conservation, sustainable use and equitable sharing of benefits, must be implemented to conserve genetic resources for our future generation.

Population Genetics for Conservation

Effective conservation of biodiversity, both ex situ and in situ, is of vital importance. Plants are immobile and possess flexible breeding systems, factors greatly influencing the appropriate methods of conservation. Studies of plant populations in space and time are essential for effective conservation. These studies include estimation of population size for in situ conservation, how best to conserve populations of endangered species, estimation of minimum viable number of population for ex situ conservation, and study of population viability (or vulnerability).

The objective of both in situ and ex situ conservation is to conserve the maximum possible number of different alleles and/or maximum possible diversity of genotypes with as small numbers of individuals as possible, whilst permitting continuous evolution. Conserved materials should be readily accessible and utilisable for breeding and research. Appropriate estimation and characterisation of genetic variation within the collection are crucial for its efficient utilisation. Understanding of the genetic structure of a population gives insight into its capacity for response to selection, both natural and artificial, and potentially its use. It also affects determination of the spatial distribution of variability and the implications of this for the collection and conservation of biodiversity.

The extent of variability within and between populations can be determined by application of both biometrics and molecular techniques. Biometrics can give us an idea of population structure in terms of means and variances for particular traits. However, the direct measurement of phenotype is time-consuming, laborious and strongly affected by environmental variations. Molecular analysis is a powerful tool which allows measurement of population structure at the gene and genome level in a way which is generally not affected by differences in environmental conditions.

DNA markers

The application of DNA-based technologies has provided a large amount of information on the population diversity of wild populations and semi-natural cultivars. Nowadays, various molecular tools are available for the characterisation of populations, e.g. isozymes electrophoresis and DNA merkwes such as restriction fragment length polymorphisms (RFLPs), randomly amplified polymorphic DNA (RAPD), amplified fragment length polymorphism (AFLP), and simple sequence repeats (SSRs).

Application of these techniques allows the study of several diverse parameters of variability within and between populations. Population studies using these techniques allow estimation of indices such as the percentage of polymorphic loci, the average and effective numbers of alleles per locus and degree of heterogeneity. Heterozygosity indices have been developed using molecular approaches. Similarly, the various measures based on the F statistics, a measure of the level of heterozygosity within and between populations have been deduced from molecular marker studies.

Population structure is controlled by several factors. These are often affected by past selection pressures which influence genetic variation both within and between populations.

Breeding system

The breeding system determines the flow of variability within a species. Sexual species, have the capacity for recombination and the resultant recombinant progenies, may be exposed to environmental selection. Outbreeding species in particular maintain a high level of heterozygosity within individuals and variability both within and between populations.

Selection, drift and gene flow

The differentiation of population structure largely depends on three processes: selection, drift and gene flow. Selection pressure represents the effect of the environment on population structures. It is influential in defining the underlying genetic structure. Random drift can lead to arbitrary fixation of genes, especially in small populations. The immigration of new genes from genetically distinctive neighbouring subpopulations can boost the selective response and enhance genetic drift by the introduction of new alleles. On the other hand, it can delay adaptation to microenvironments by repetitive introduction of alleles adapted to a different neighbouring microenvironment.

Forage Species in Agriculture

Forage legumes are essential to livestock industries across the world and they play a major role in high quality grazing and conserved feed for the efficient production of meat and dairy products. They also contribute significantly to soil fertility and conservation through nitrogen fixation. Most forages, especially legumes, are multi-purpose plants. Apart from livestock products they can also contribute to weed control and pest and disease reduction on other crops. Forages can also be used as green manures, improved fallows, cover crops and live fences. Their capacity for nitrogen fixation and high protein content reduces the need for external inputs of fertilisers and concentrate feeds. The high feed quality of forages legumes leads to improved manure and compost with further beneficial consequences for livestock production.

In 2003, the total area of the UK agricultural land was 18.4 million hectares, over 70 per cent of the total land area. The total value of production of forage plants in UK was £163 million. This included: peas, beans and maize grown for stockfeed, farm transfer of peas and beans, hay and dried grass, grass for grazing livestock and non-agricultural horse grazing. Approximately 70 per cent of the UK total agricultural land is used as grassland and rough grazing.

White clover (*Trifolium repens* L.) is often cultivated within its natural range, for instance in the UK more than 11 million hectares of agricultural land is under pasture and it frequently consists of *Lolium. perenne* and *T. repens*.

The genus Trifolium L. contains about 250-300 species and it is a member of the Fabaceae (Leguminosae) family, subfamily papilionaceae. *Trifolium* is grouped with closely related genera, *Trigonella, Medicago,* and *Melilotus,* in tribe Trifolieae. The genus *Trifolium* includes more than 20 clover cultivated species as forages. About 54 wild species are only distributed in North America and about 33 exclusively south of the Sahara in Africa. The remainder occur mostly in Europe, Northern, south-western Asia, and North Africa. The Mediterranean region is the most concentrated area of distribution and about 40 species are confined practically to the Orient region (Turkey, Syria, Cyprus, Lebanon, Jordan, Israel, Iran and Iraq) (Hossain, 1961 cited by Williams 1987a).

Morphological Features of White Clover

The taxonomic description of *T. repens* L. is as follows; Perennial, globrous or glabrescent, 10-30 cm.

Figure 2. Illustration of Trifolium repens L. in flower 1: main stolon, 2: inflorescence, 3: leaf, 4: petiole, 5: root

Stems rhizomatous, prostrate, long-creeping and rooting from nodes. Leaves long-petioled, sometimes petioles hairly; stipules broad at base, ovate-lanceolate, with subulate upper part, scarious with reddish or green nerves; leaflets 0.6-2.5(-4) × 1-1.5 cm, broadly obovate to orbicular, obcordate, mostly emarginate, margin sharply serrulate, with parallel, lateral veins forked towards margin.

Inflorescences 1.5-3.5 cm in diam., umbellate, 20-to many-flowered, rather loose, nearly globular. Peduncles as long as or much longer than subtending leaves, weak. Bracts shorter than pedicels, ovate-oblong, acuminate. Pedicels as long as or longer than calyx tube, somewhat deflexed from early flowering, strongly deflexed after anthesis, glabrous or hairy. Calyx 3-5 mm, glabrous; tube campanulate, 6-10 -nerved teeth about as long as tube, somewhat unequal, triangular-lanceolate.

Corolla 4-13 mm long, white, yellow or pink; standard ovate-lanceolate, oblong, rounded at apex; wings somewhat apreading. Pod 4-5 mm, usually 3-4 seeded, linear-oblong, constricted between seeds. Seeds ovoid to reniform, brownish. Flowering time March-September. 2n=16, 28, 32 48, 64.

Habitat

T. repens grows in damp and swampy soils and also on lawns and grassy places. It is tolerant of moderately acidic to moderately alkaline soils, wet soils, and flooding but does not grow well under saline conditions or heat and drought. Life cycle: Perennial.

Reproduction and Dispersal

White clover is an outcrossing species pollinated by honey bees and bumble bees. Long distance seed dispersal is mainly through the digestive tracts of birds and grazing animals. Seeds can be spread over short distances by dehiscence, by tramping, by earthworms and to a small extent by wind. Seeds are easily shaken out when it is ripe. The species also spreads robustly by stolons (horizontal creeping stem).

Invasive Potential

Spreads vigorously by stolons and reseeds naturally when it matures. Because of its progressive production of seedheads, large numbers of seed are produced. In wet conditions, it may spontaneously appear in early spring in thin plant stands or on bare soil where suppression is weak or absent. It is a diminutive plant that can be readily shut out by other plants especially as soil nitrogen status improves.

Evolution of Species

T. repens is an allotetraploid species of chromosome constitution 2n=4x=32, and it contains a high level of genetic heterogeneity. Polyploidy is uncommon in Trifolium and has not been a major mechanism of speciation in the genus. The evolutionary origins and progenitor of white clover are not fully elucidated, although two diploid species (*T. occidentale D. Coombe* and *T. nigrescens* Viv.) and another allotetraploid species (*T. uniflorum*) are considered as close relatives due to the relative ease with which interspecific hybrids may be produced.

There are also close similarities amongst the karyotypes of *T. repens, T. nigrescens, T. occidentale,* and *T. ambiguum.* Considering their close relationship and nature as widespread diploid species, *T. repens* might have arisen from a cross between *T. nigrescens* and *T. occidentale.*

Geographic Distribution of White Clover

Even though it is considered as a temperate species; the geographic distribution of white clover is exceptionally broad. It is found where soil moisture is adequate for growth. White clover is adapted to a wide climate range from the Arctic to the subtropics and has a wide altitudinal ranges. It is found as high up as 6000m in the Himalaya Range of India. The likely centre of origin of white clover is the Mediterranean region. The indigenous area includes the whole of Europe and central Asia, west of Lake Baikal as well as areas in North Africa. (Morocco, Tunisia). It has become naturalised in China, Mongolia, Korea, and Japan. It was also introduced in north, South America and Australia.

T. repens is a highly variable species and considerable differences have been observed in a wide range of morphological characters. The wild type, *T. repens* var. repens, occurs throughout the entire distribution of the species. Varieties of white clover may be grouped into small (common or wild), intermediate ('Dutch' and 'New Zealand'), and giant (ladino) types. Williams classified major ecotypes of white clovers as follows;

Ladino Clover (Italy)

These populations are consistent with the hypothesis that the Mediterranean is a centre of diversity for the species because of a wide range of variation in Italian white clover populations. The most important representative of this ecotype is Ladino gigante Lodigiano. All of its organs are very large and intolerant of frequent grazing. The area of origin of this ecotype is northern Italy, which has relatively cold winters with mean January temperatures of around 0°C. Ladino clover is not well adapted to sustained drought.

Morsø and Strynø

Two Danish islands were the region of origin of these ecotypes. Morsø is a wild clover from the island of Mors in the Limfjord north of Jutland. It comes from excellent old permanent pastures, which were persistent and somewhat coarser than English wild white. It now forms the basis of northern European bred varieties.

Dutch White Clover (Western Europe)

The name was given by English farmers in the early 20th century. It means all non-wild strains or cultivated white clovers. The term is now little used, although they are still characteristics of Dutch white clover can still be seen in some European cultivars. Its name implies that it was first cultivated in Netherlands but the origin is unknown. Dutch white clover is found in open habitats and has medium-sized leaves profuse early inflorescence.

English Wild White Clover

This extremely hardy and long-lived clover was popular in the 19th century. It has prostrate, small-leaved, dense form character. Its small leaves, thin stems and short petioles are recognised as an ecological adaptation to very old, constantly grazed sheep pastures in the United Kingdom. It shows that highest productivity under close grazing condition and great persistence. It is also more tolerant of low fertility soils than larger-leaved clovers.

New Zealand Ecotypes

White clover was introduced into New Zealand in the mid 19th century by immigrants and by the 1920s a lot of variation within and between the populations had developed. There are six major types; New Zealand wild white Number 1, New Zealand wild white Number 2, Ordinary New Zealand white, Lax early-flowering New Zealand and ordinary European white (New Zealand 'Dutch') , Small-leaved Kentish types, and Large leaved Ladino types.

Australian Ecotypes

Just as in New Zealand, white clover does not naturally occur in Australia. It was introduced by early settlers and ecotypic differentiation has occurred in a short period of time. 'Victorian Irrigation' was developed in Northern Victoria where the summer is very hot and dry and is particularly suited to irrigated grassland. In the Clarence Valley in New South Wales, a naturalised ecotype occurs but this is not widely seen every year.

Ecotypes in the USA

North America is not an area where white clover occurs indigenously: Nonetheless, natural populations are widespread and accompanying ecotypes are differentiated. There is a report that white clover already existed in USA in 1775, and it is thought that the French introduced it in Canada around 1700. The ideal conditions of the region made the species spread quickly. The most distinct USA ecotype has been improved to become more winter hardy. Natural populations from New Jersey proved to be enormously variable.

Subtropics

White clover is not adapted to the subtropical climate. However, a considerable number of populations were introduced in high-altitude areas of Kenya, Colombia, Hawaii, sub-tropical Australia, and parts of India. It is necessary for white clovers which occur in those regions to adapt to day length, lack of defined seasons, large fluctuations between day and night temperature and lingering hot dry periods. Special tolerance to high temperature and drought is also required.

Genetic Diversity in White Clover Populations

Although the primary mechanism for growth is

vegetative propagation or clonal spread, white clover populations are genetically highly variable. Studies using morphological and genetic techniques have showed that white clover populations in permanent grassland are characterised by remarkable genetic diversity within very small areas. Burdon argued that highly variable microenvironments select individuals which are best adapted genetically to a given set of conditions, such as age, size of neighbouring plants, sheep grazing, etc.

Thus, the variable environment exerts a strong disruptive selection force on the clover population and maintains high genetic diversity. In addition, Gustine concluded that both commercial cultivars and indigenous ecotypes readily adapt to dramatically changing environments short periods (two months) through differential growth among individuals and response to selection pressure.

Economic Importance of White Clover

White clover is one of the most important forage crops for ruminants, particularly sheep and cattle. Cultivars that are well adapted to specific environments and management conditions may significantly enhance pasture production systems. The development of sustainable farming practices and the growth of the organic farming sector have all contributed to increased use of white clover due to its excellent nutritive value as forage and its and ability to fix atmospheric nitrogen improving the quality of soil.

Nitrogen fixation of white clover reduces or eliminates requirement for input nitrogen fertiliser. Use of clover can result in significant financial benefits for farmers by reduction in costs and better individual animal performance. Moreover, from the ecological point of view, nitrate leaching into water causes important for preventing eutrophication. Besides its agricultural importance, white clover also has potential ornamental value.

Forage Quality

The inclusion of clover in a sward improves forage quality and consequently value for feeding animals for several reasons, in particular digestibility and minerals and protein contents (Table 1).

Table 1. Typical quality characteristics of white clover and perennial ryegrass.

	White clover	Perennial ryegrass
Digestibility (Range) (% D value)	75-82	65-75
Crude protein (%)	27	17
Amount eaten by sheep (Dry matter intake, kg/day)	1.9	1.4
Mineral contents		
Calcium (%)	1.6	0.6
Magnesium (%)	0.18	0.16
Phosphorous (%)	0.6	0.3
Copper (ppm*)	10.0	6.5
Selenium (ppm* ×100)	0.6	0.2

* Parts per million

Digestibility of clover is high and it has considerably greater protein content than ryegrass. Mineral contents is also generally higher than that of grasses. The production of nutritionally compatible grass varieties (high sugar) may allow more efficient results of utilisation of clover protein.

Nitrogen Fixation

In symbiosis with Rhizobium bacteria, white clover can

fix nitrogen from the air and convert it to a form which can be utilised by the plant. In the development of symbiosis, Rhizobium leguminosarum biovar. trifolii proliferates and initiates infection of the roots of white clover, forming small growth of tissue, called nodules. Atmospheric dinitrogen gas is reduced to ammonia within the nodule by the nitrogenase enzyme system. Subsequently, the nitrogen is transferred into soil in different ways; death and decay of old plant part such as roots and nodules, grazing animals' dung and urine, and ploughing.

Breeding Targets

In breeding of white clover, the main objective is to improve production from grazing animals. White clover is largely grown together with a companion grass such as ryegrass (*Lolium perenne L.*). Therefore, production of balanced swards with appropriate white clover contribution is the key concern of white clover breeding in the UK. Persistence, compatibility with Rhizobium bacteria and compatibility with insect pollinators are also important elements in the agronomic success of white clover. Tolerances of abiotic: low temperature, drought and acid soil, and biotic: slug, nematode, virus and fungal pathogens, stresses are target of white clover breeding programme around the world.

17

Bi-cropping System for Growing Cereals

Temperate legumes, grown in association with grasses, can fix over 250 kg of N ha^{-1} annually. On average, about 100kg ha^{-1} is harvested directly as legume herbage, 50 kg ha^{-1} indirectly as extra grass and the remaining 100 kg accumulates as organic N in the soil. Usually the realization of this organic N is dependent on ploughing and arable cropping; its value is reflected by the reduced requirements of the subsequent crops for fertilizer N.

The development of drills capable of introducing seed into undisturbed soil made it possible to sow a non-leguminous species in an established legume sward. In such a system the non-legume might benefit form both residual and current nitrogen fixation. Survival of the legume understorey would lead to the process being repeated with a succession of crops.

Growth of Novel System

A system is being developed at IGER to grow cereals, especially winter wheat, in a permanent understorey of white clover. The system is straightforward. At first a pure sward of white clover is established, given a year's growth to build up fertility and then grazed hard during the autumn. The cereal crop is then direct-drilled into the clover. A standard Hunter Rotaseeder is particularly suited for this purpose. The cereal is grown with minimal or now nitrogen fertiliser and harvested conventionally. After harvest the clover is allowed to recover, grazed down hard and re-drilled with cereal in the autumn or spring to repeat the cycle.

The continuous presence of a crop should minimise the loss of nitrogen through leaching because the roots of cereal and clover absorb any mineralised nitrogen. The presence of leaf area for the full period of growth maximises the use of solar energy. Establishing cereals successfully and maintaining the desired balance of components to ensure reasonable cereal yield and clover survival is a skilful operation (Table 1). Cereal establishment was assured by suppressing the clover through grazing heavily with sheep.

Clover survival was aided by choosing cereal varieties with short, stiff straw and drilling in wider than normal rows. This reduced the risk of lodging and also allowed some light to penetrate the cereal canopy. Clover recovery was immediate from existing growing points.

Table 1. Yield of grain (t/ha, 15% moisture) of barley and oats grown with and without N, either conventionally or in a clover understorey

Treatment	Barley		Oats	
	Zero N+	100kg N/ha	Zero N+	100k N/ha
Ploughed/conventional seedbed	5.96	5.19*	4.80	3.88*
Clover understorey	5.66	5.31	5.88	3.55*

* crop lodged

Choice of Cereal

Trials with a range of cereals showed that winter wheat

and spring barley appeared to be the most compatible for maintaining the clover sward. However spring barley may require agrochemical suppression of the clover for good establishment. Winter barley tillered profusely and became too competitive, even at modest seed rates and wide drills. Spring wheat and spring oats produced few tillers allowing the clover to become dominant. Winter oats are too tall and intercept nearly all the light for long periods, leading to the clover dying. Winter wheat, since it tillers less, is best suited to the system. The density of the wheat canopy can be manipulated more easily than other cereals by altering seed rates.

Pests, Diseases and Weeds

Pest and disease problems to date using the clover understorey system have been much reduced, although the reasons for this are not clear. Populations of aphids were low or below the level of detection in plots where the clover was present and did not need control with insecticides. By contrast an adjacent area, which was grown conventionally, received several sprays.

Although there is no objective evidence, we believe that the dense year-round cover of vegetation provided by the clover offers a refuge for many species of predatory invertebrates. The most important of these are probably spiders, carabid and staphylinid beetles which feed on cereal aphids. Also it is known that alighting cereal aphids tend to select particular patterns of vegetation. The continuous ground cover achieved with the system could reduce the number of aphids colonising the cereal crop.

Cereal diseases have not been a problem. This may be partly because we used seed dressed with fungicides and we selected disease resistant varieties. Also, the wide row spacing and the more open crop canopy probably altered the microclimate to make it less suitable for disease development.

So far, weeds have not been a problem. But they are likely to be more troublesome than in a conventional crop because only a limited number of herbicides are available to control broad-leaved weeds without damaging the clover. Prevention is better than cure and keeping a good canopy to prevent weed invasion seems to be the best solution.

18

Nutritive White Clover

White Clover Content

The white clover content in the sward is obviously very variable, depending on many factors such as fertilisation, grazing or cutting, grazing management etc. In the regrowth after defoliation, however, the white clover content seems to be constant.

A nearly constant clover content above stubble level after re-establishment is also shown in Fig. 1. The overtopping of the clover in the last part of the regrowth period does not seem to affect the clover content.

The typical progress with the highest clover content in the middle of the growing season at invariable treatment is also shown in the figure. The defoliated part of white clover consists of 'leaves' (leaves and petioles) and 'flowers' (inflorescences and flowering stems).

The stolons can only be defoliated to a reduced extent, e.g. at sheep grazing. The flowering takes place primarily in the middle of the growing season, and there are only very few flowers in spring and autumn. In mid summer the 'flower' fraction can attain 50 % of the white clover (Fig. 1).

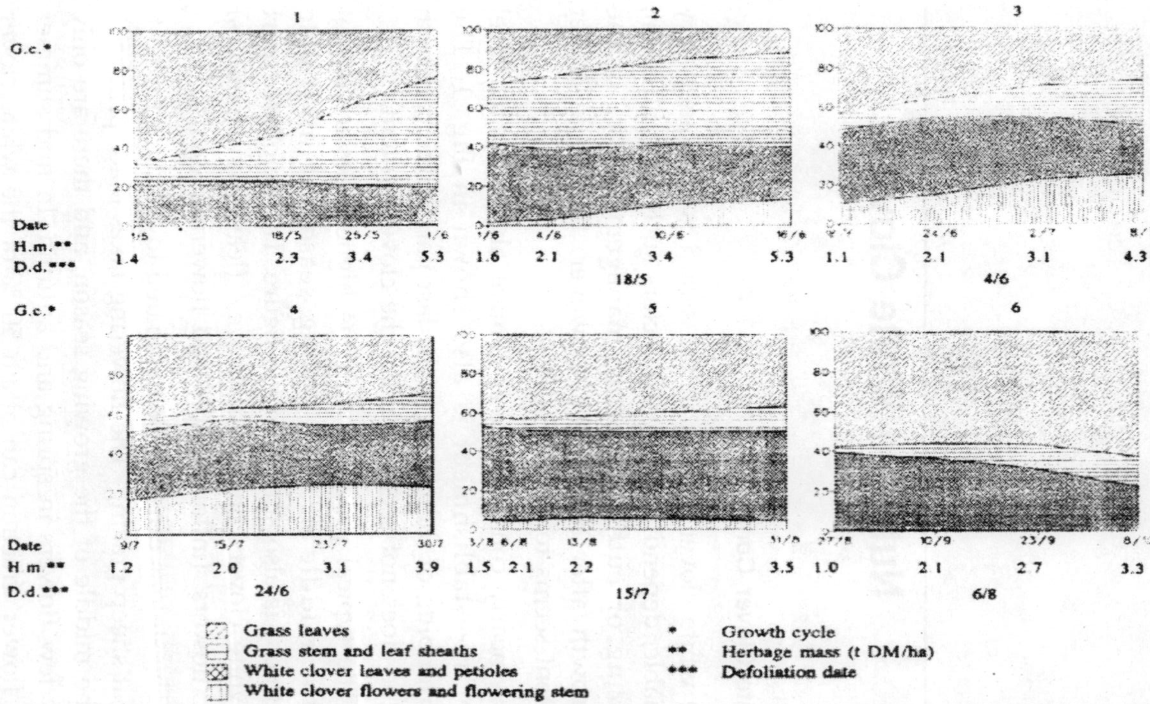

Figure 1. Botanical composition in a sward of perennial ryegrass and white clover, simulated grazing (defoliation at 2.0 t DM/ha), irrigated, 1 harvest year 1992.

Digestibility

The digestibility of pure grass compared to grass mixed with white clover is reported differently, e.g. decreasing or unaffected digestibility when white clover is added. How white clover affects the digestibility of the sward must depend very much on the conditions, the grass species, the defoliation management etc.

When comparing at the same herbage mass, with a relatively large number of samples, no effect on digestibility was found by mixing grass with white clover. The clover content was 25-45 %. The relationship between the digestibility and the herbage mass was the same for the two different crops in the single growth cycle. At 2 t DM/ha above stubble level the IVOMD was 85, 82, 79, 76 for 1., 2., 3., 4. growth cycle respectively. The content of grass stem was increased by addition of white clover. This may affect the digestibility of the grass negatively. The effect was greatest in the mid summer, when the clover content was highest.

Table 1. Chemical composition of plant fractions in perennial ryegrass mixed with white clover fertilised with 200 kg N/ha. 25-40 % clover of DM. 1991, 17 obs. mean of the growing season.

	Perennial ryegrass		White clover	$LSD_{0.05}$
	Leaves	Stems and sheaths		
N (% of DM)	3.1	1.9	4.1	0.3
NDF (% of DM)	37.8	43.3	26.2	3.6
ADF (% of DM)	23.7	26.6	22.5	6.6
Lignin (% of DM)	1.7	2.1	3.9	0.6
IVOMD	78.0	73.8	77.3	2.6

In the same mixed sward there was no difference in the digestibility of the grass leaves and the white clover (Table 1). However, the clover consisted of both 'leaves' and 'flowers'. Gibb and Theacher and Wilman and Altimimi found a low digestibility in the flowering stem, and especially in the flower. The leaves and the petioles had the highest digestibility.

Corresponding results are shown in Fig. 2. The digestibility of the clover 'flowers' was lower than the grass 'stem' and the digestibility of the clover 'leaves' was higher than the grass leaves.

It is often pointed out that white clover maintains a high digestibility in the spring. This is also confirmed by the results in Figs. 1 and 2, because there were no 'flowers' in the spring growth, and the digestibility of the clover 'leaves' maintained a high level. But over the largest part of the summer the clover is flowering and the 'flowers' lead to a lower digestibility. As the 'flower' content increases during the regrowth period (Pig. 1), this also explains why the digestibility does not increase by addition of white clover in the regrowth periods.

Cell Wall Composition

Normally white clover has a considerably lower content of total cell wall or neutral detergent fibre (NDF) than grass and a higher content of lignin. The lowest NDF content was found in the leaves of white clover. In the petiole the content was higher, whereas it increased in the flowers with advancing maturity. The content of cellulose or the content of acid detergent fibre (ADF) can in some cases be the same. At the same herbage mass addition of white clover did not affect the ADF-content either.

As a consequence of the lower content of NDF and the same digestibility when comparing at the same herbage mass, it is tantamount to a lower digestibility of the clover cell wall. Pectin is not, however, included in

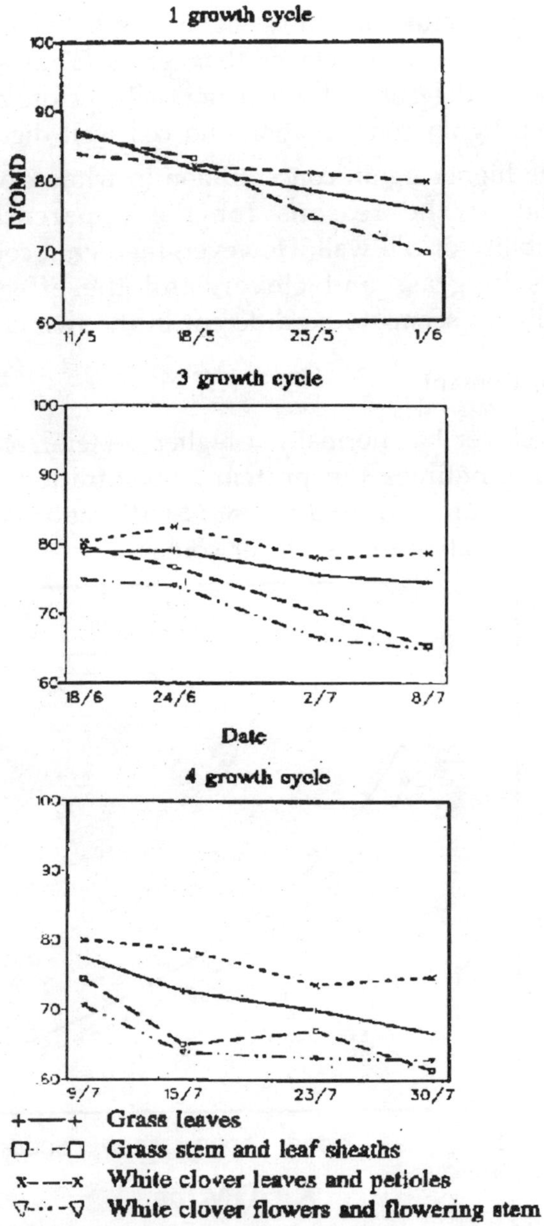

Figure 2. In vitro organic matter digestibility of different plant fractions. Corresponds to the results shown in Figure 1.

the NDF fraction, but it is highly digestible. The pectin content is higher in clover than grass (Thomson, 1984) and not so digestible. There is normally a close agreement between lignin concentration and cell wall digestibility.

The higher lignin concentration in white clover could be one of the reasons for the apparently lower digestibility of cell wall. However, the lignin composition differs in grass and clover, and the effect on the digestibility seems to be different in the two crops.

Protein Content

White clover has normally a higher protein content than grass. Sometimes the protein concentration in mixed crops is unaffected by different N-fertilisation on account of compensation of the clover content.

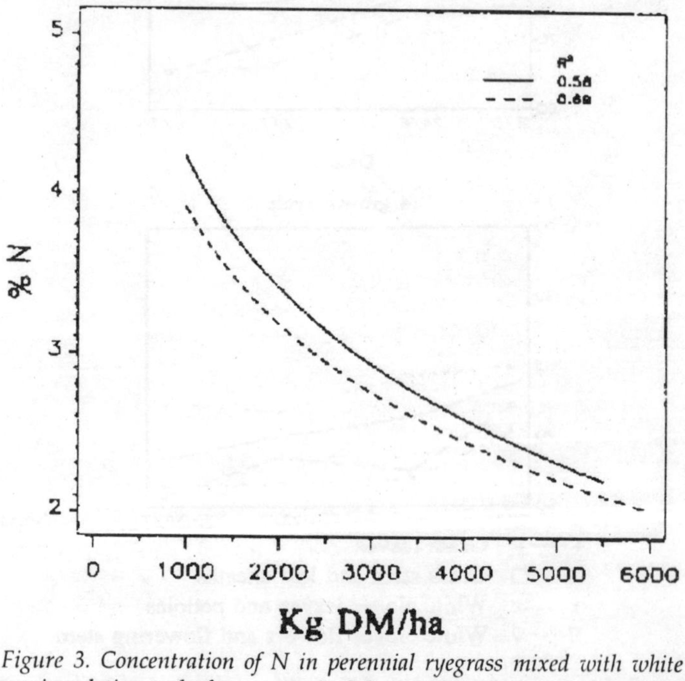

Figure 3. Concentration of N in perennial ryegrass mixed with white clover in relation to herbage mass in 2-3 growth cycle (-----) and the other (1, 4, 5, 6) growth cycles (- - - - -), 1989-91.

In grass grown with white clover the protein concentration can also be increased compared to grass grown without clover. The N content decreases with increasing herbage mass (Fig. 3). At the same herbage mass the N content was a little higher in the mid summer, where the clover content was highest (Fig. 3).

The higher protein content in clover does not seem highly digestible in the rumen. Sanderson and Wedin found that concentration of N in ADF in clover was twice the concentration in grass, which may be due to the greater lignin concentration. Maybe this is one of the reasons for relatively higher N digestion in the intestine in sheep fed with white clover than in sheep fed with perennial ryegrass.

Content of Minerals

The concentration of some of the minerals are normally higher in white clover than in grass. It varies which minerals are most influenced. In one experiment e.g., the concentration of K and Ca increased, whereas Mg, Na and P decreased. In another experiment the concentration of both P, Ca and K increased.

Table 2. Chemical composition and clover content (% of DM) in pure perennial ryegrass (G) and perennial ryegrass mixed with white clover (CG). Sandy soil, mean of the years 1985-87.

	kg N/ha	Clover content	N	K	Ca	Me	P
CG	0	75	3.57	3.57	1.18	0.24	0.39
CG	175	37	2.70	3.48	0.86	0.21	0.40
CG	300	16	2.49	3.49	0.70	0.20	0.41
CG	425	8	2.61	3.51	0.68	0.20	0.41
G	425	-	2.34	3.49	0.58	0.19	0.40
$LSD_{0.05}$		6	0.15	0.08	0.07	0.01	0.01

In Table 2 results from a fertilisation experiment are shown. The concentration of Ca and Mg was the most positively affected by decreasing N fertilisation and by this increasing clover content. On the other hand the K concentration was only slightly affected. In this case the risk of grass tetany decreased by addition of clover in the sward.

When experiments show different results concerning effects on minerals, it may be due to the fact, that many parameters affect the concentration of minerals, e.g. temperature, availability, soil type and application.

Intake

White clover addition to grass increases the intake by ruminants both at silage feeding and under grazing even at the same digestibilities. This fact has been accepted for a number of years. Thus white clover has a lower retention time in the rumen, and a relatively higher amount enters the small intestine. The apparent digestion can therefore be reduced with white clover in the diet. The higher rate of passage is especially explained by the lower content of cell wall or structural fibre, lesser resistance to mechanical breakdown by chewing and, in addition, an increasing amount of viable bacterial populations in the rumen.

19

Animal Responses to White Clover

The continuing overproduction of animal products within the European countries has stimulated interest in extensive management systems. As stocking rates are reduced the opportunity for incorporating white clover in the production cycle increase. There is also concern over the importation of nutrients and resources into intensive livestock systems and the subsequent problems associated with inefficient utilisation of nutrients, nitrogen especially, and the disposal of the waste products.

Extensive livestock systems generally rely upon reduced variable costs, such as purchased fertiliser nitrogen, to offset any reductions in financial output that occur. However, if improved animal performance can also be achieved as a consequence of including white clover in the diet, the financial viability of the system will be further improved. Thomson suggested that the daily liveweight gain (DLWG) of sheep grazed on grass/white clover swards could be 25-30% higher than those on conventional grass swards, and 15 -20% for beef cattle, while a possible increase of 10% was suggested for milk production from grass/white clover swards. The following chapter reviews some of the responses that have been observed when clover has been incorporated in diets.

Nutritional Value

White clover has a higher nutritive value than grass, due largely to its lower structural fibre and higher protein content. These characteristics give white clover improved digestion and higher intake capabilities. Beever *et al.* showed that growing cattle maintained on ryegrass from May to September had mean intakes of 21.8 g/kg liveweight compared to 27.0 g/kg for clover. It has been observed that the rate of dry matter (DM) degradation during the first three hours after feeding can be 15% higher when fed clover compared to ryegrass. They associated this with a greater reduction in particle size as twelve hours after feeding on clover the ratio of small: large particles in the rumen (< or > than 1 mm) was 3: 1, compared to 1: 1 when fed on ryegrass.

Beever *et al.* observed even greater differences in rates of DM degradation in cattle with 14%/h occurring on a clover diet compared to 7%/h for grass. They concluded that the extent of DM digestion observed on a white clover diet after 8 hours would not be achieved until 14 hours post feeding on a grass diet.

Steg *et al.* compared *in situ* ruminal breakdown and intestinal protein digestion in dairy cows of N-fertilized perennial ryegrass (315 kg N/ha) and the white clover and ryegrass components of a mixed grass/white clover sward, fertilized with 90 kg N/ha in spring. They observed higher soluble fractions of organic matter (OM) and crude protein (CP) and lower rumen undegradable fraction of OM and CP in clover than in grass. The degradation rates of OM and CP were higher and similar respectively to those of grass.

Increased intakes of silage DM have also been observed when clover was incorporated in the ration. Castle *et al.* (1983) mixed pure ryegrass and pure clover silage in differing proportions and observed that 50% inclusion rate of white clover silage raised the total silage dry matter intake (DMI) of early lactation dairy cows by 13% compared to offering only grass silage. Offering 100% clover silage raised the silage DMI by 17.5%.

The inclusion of clover in the diet can also promote changes in the nitrogen metabolism of ruminants with the potential for considerable losses of nutrients. Beever *et al.* fed either fresh ryegrass or clover to cattle at three stages through the growing season. The increased crude protein content of the clover herbage compared to the ryegrass resulted in large increases in mean rumen ammonia concentrations (Table 1).

However, despite a 90% increase in dietary crude protein intake the average duodenal non ammonia nitrogen (NAN) supply was only 15% higher when the cattle were offered clover. In order to compensate for the increased losses of N in the rumen Steg *et. al.* suggested that clover based rations be supplemented with rumen-degradable carbohydrates.

Table 1. Nitrogen metabolism of cattle fed fresh ryegrass or white clover

Cut	early	Rye grass mid	late	early	White clover mid	late
Dietary crude protein (g/kg OMI)	160	154	190	318	307	329
Mean rumen ammonia cone (mg/1)	50	68	242	283	372	393
Duodenal NAN (g/kg DOMI)	41.9	37.2	39.0	40.1	50.8	44.8

Dairy Cattle

The production responses observed in dairy cattle can be related to the clover content of the diet. In an experiment reported by Thomson *et al.* spring calved cows were grazed either on pure perennial ryegrass or pure white clover swards from weeks 4 to 18 of lactation. During this period the clover grazed cows produced 2.8 l/d more milk (Table 2).

Table 2. The productivity of cows grazing either perennial ryegrass or white clover (lactation weeks 4-18)

	Perennial ryegrass	White clover
Milk yield (l/d)	22.2	25.0
Milk fat (g/kg)	29.8	30.9
Milk protein (g/kg)	41.5	38.9
Total milk yield (1)	2396	2697

In addition to the improved milk yield and increased yields of fat and protein, they noted that the grass fed cows suffered a greater and longer period of body weight loss before repletion commenced at day 102 post calving compared to day 77 for the clover fed cows.

Further studies with cows grazed on mixed swards have demonstrated the potential benefits from white clover. Wilkins *et al*. grazed spring calving cows on swards that contained either 0.01 (1), 0.16 (m) or 0.23 (h) of the above ground herbage as white clover (Table 3).

The experiment resulted in large differences between milk yield on the three sward types. When no concentrate supplementation was offered the fat and protein corrected milk yield (FPCM) was 5.6 kg/d higher from the high clover sward compared to the low clover sward. It was noted however that the nitrogen supply from the low clover sward may have been limiting as none of the swards received any nitrogen fertilizer. Of more interest was the increase in yield of 3.4 kg FPCM/d when cows were offered the high clover swards compared to the medium, which at 16% clover content would be typical of better grass/white clover swards that cows are turned out onto in the spring. Supplementation with concentrate produced similar responses in milk yield in the cows grazed on both the medium and high clover content swards.

Table 3. Milk production and composition for grazed swards with contrasting white clover contents (April-July)

		Yield		Composition (g/kg milk)		
Clover content	concen trate (kg/d)	kg/d	FPCM (kg/d)	Fat	Protein	Lactose
Low	0	20.6	18.8	35.9	26.8	41.2
	2	23.2	21.4	36.4	27.6	42.1
	4	25.5	24.4	38.6	28.4	43.2
Medium	0	22.8	21.0	35.8	28.2	42.0
	2	27.1	25.7	38.1	28.4	42.8
	4	26.2	25.2	38.3	30.2	43.3
High	0	25.4	24.4	39.1	28.6	41.9
	2	25.9	25.0	39.0	29.5	43.6
	4	26.3	25.5	39.1	29.7	43.1

Preliminary results from ongoing experiments in The Netherlands are presented in Table 4. Bruins compared the intake of dairy cows fed fresh cut grass (300 kg N/ha/year) or grass/white clover (50 kg N/ha/year in spring) with 60-65 % of white clover in the DM. The herbage fed in May was from a second cut with a growing period of two to three weeks and the herbage fed in autumn was from the fifth and sixth cut with a growing period of four to five weeks.

Table 4. The intake and productivity of cows fed grass (G) or grass/clover (GC) (lactation weeks 18-21)

	May		August/September	
	G	GC	G	GC
Intake Herbage (kg DM)	16.5	17.5	18.5	17.6
Concentrate (kg DM)	2.3	2.3	3.8	3.8
Milk (kg/cow/d)	26.0	26.3	24.1	25.4
Fat (g/kg)	43.5	42.7	43.2	40.9
Protein (g/kg)	32.1	32.5	33.1	33.1

When the potential of grass/white clover swards has been investigated in system studies, contrasting results have been obtained. Ryan reported on a 5-year experiment in which spring calving cows were rotationally grazed on either conventionally fertilized grass swards or on grass/white clover swards (Table 5).

Table 5. Stocking rate and summer milk yield for four grazing systems

	Stocking rate		Milk yield
	(cows/ha)	(l/cow)	(l/ha)
Grass/white clover	2.08 (L)	2810	5856
Grass/white clover	2.52 (M)	2738	6908
Grass plus nitrogen	2.46 (M)	2740	6764
Grass plus nitrogen	3.20 (H)	2578	8266

Although the greatest output of milk per hectare was achieved from the grass plus nitrogen (H) treatment, the highest yield per cow was produced from the grass/white clover swards stocked at 2.08 cows/ha. The clover cows which were stocked moderately, produced a similar milk yield per cow and 144 l/ha more milk than the moderately stocked grass plus nitrogen group.

Detailed whole farm system studies investigating the potential of clover—rich swards for dairy production are being undertaken at two sites in Europe, one near Dumfries in south-west Scotland (SAC) and another at Lelystad in The Netherlands (PR). In Scotland two herds of 70 autumn—calving Friesian Holstein dairy cows plus their youngstock were maintained on either grass/nitrogen swards, receiving up to 350 kg N/ha/annum or on grass/white clover swards.

For the first two years of the study both herds were each allocated 36 ha of grassland, and 400, 000 1 milk quota. The cows were set-stocked during the summer, and offered a grass or grass/white clover silage based complete mix during the winter. It was found that between 25% and 33% more concentrate per litre of milk sold was necessary to reach the quota target on the clover based system when the overall stocking rate was maintained at 2.4 lu/ha.

Table 6. Comparison of a grass/nitrogen system with a grass/clover system on animal performance when managed at similar stocking rates

	Grass/Clover		Grass/Nitrogen	
	1988/89	1989/90	1988/89	1989/90
No cows	71	70	70	72
Milk sales (l/cow)	5658	5605	5764	5532
Cones (kg/cow)	1709	1554	1412	1185
Milk fat (g/kg)	40.0	39.9	39.5	40.7
Milk protein (g/kg)	31.9	31.9	31.8	31.3

Subsequently, the intensity of production was reduced when the land area of the grass/clover herd was increased by 25% which reduced the overall stocking rate to 1.9 lu/ha. The reduction in stocking rate for the grass/clover herd ensured that there was no shortage of forage and when both herds were allocated the same milk quota, the concentrate use, milk composition and milk yield per cow were very similar (Table 7).

Table 7. Comparison of a grass/nitrogen system with an extensive grass/ white clover system

	Extensive grass/white clover		Grass/Nitrogen	
	1990/91	1991/92	1990/91	1991/92
Milk quota	420,000	350,000	400,000	400,000
No of cows	70	70	69	69
Milk sales (l/cow)	5719	5294	5724	5941
Conc use (kg/cow)	1096	501	1101	1077
Milk fat (g/kg)	40.5	41.3	40.4	40.8
Milk protein (g/kg)	32.4	31.8	31.8	32.0

However, there was no production response to the white clover, with the more extensively managed herd yielding only 22 kg FPCM more per cow than the herd maintained on conventional grass swards. In the following year the target milk yield for the extensive herd was reduced to 5000 l/cow, and as a consequence the concentrate use dropped to 501 kg/cow. The increase in milk fat and reduction in milk protein could be attributed to the increased forage: concentrate ratio rather than any clover effect.

The lack of animal response over the four—year period may have been partly a function of the herbage quality and of the calving pattern. The crude protein content of the grass/white clover herbage was consistently lower in early season than that of the grass/ nitrogen swards, which also resulted in low crude protein first cut silage which provided the bulk of the winter feeding. The crude protein ranged from 126 to 107 g/kg DM with corresponding clover contents of 12.5 to 17.1% in total dry matter.

However, in year three, the crude protein content of the grass/white clover silage was 171 g/kg DM. This may have been due to cutting earlier and an increased clover content of 28.1 %. The clover contents of the mixed swards followed the normal pattern, with mean values of 39.3% and 38.8% at second and third silage cuts respectively over a four—year period. Both herds calved between October and December each year, and at turnout the following spring were in mid-lactation (lactation weeks 16-29). The lack of animal response could have been due to a winter feeding regime based on silage with relatively low clover and crude protein contents, followed by turnout onto pastures in which nitrogen could have been limiting animal responses in early season.

Table 8. Comparison of a grass/nitrogen system with a grass/while clover system

	Grass/Clover		Grass/Nitrogen	
	1990/91	1991/92	1990/91	1991/92
Milk quota (1)	420,000	420,000	420,000	420,000
No of cows	57	58	58	58
Milk sales (1)	7402	7174	7180	7177
Cone use (kg/cow)	1599	1675	1582	1648
Milk fat (g/kg)	45.1	45.6	44.6	45.1
Milk protein (g/kg)	34.2	34.8	34.5	34.8

The systems study in The Netherlands consisted of two herds of 60 winter (October-April) calving cows plus 40 youngstock. The grass/nitrogen swards received approximately 275 kg N/ha and the grass/clover swards were given 70 kg N/ha in spring. With an anticipated yield difference of 20% the area of the grass/clover and nitrogen farm was 41 and 34 ha respectively. Dairy cows and youngstock grazed rotationally on paddocks of 1.25 ha in a leader/follower system.

Silage was cut at an average DM yield of 2500-3000 kg/ha. There were only minor differences in energy values in silage of both systems. The average (1990/91 and 1991/92) crude protein content of grass/clover and grass/nitrogen silages were 175 and 166 g/kg DM respectively. The results in 1990/91 showed a minor advantage in milk production for the grass/clover system but in 1991/92 there was no difference at all. The concentrate use in 1991/92 was higher because of extremely dry weather in August and September of that year.

The average white clover content (October) was satisfactory with 34 and 18% respectively, but there was a large variation between the paddocks with a minimum of 1 % and a maximum of 68%. This variation was partly a result of variation in the initial phase after seeding but also a result of other factors such as management and grass and clover variety. The consequences of this high variability were paddocks of poorer quality with low clover contents and an increased incidence of bloat where there were higher clover contents.

Beef Cattle and Sheep

In beef systems enhanced rates of liveweight gain have been reported on grass/clover swards and silage, but as with dairy cattle a response does not always occur. Younie *et al.* compared Hereford and Friesian steers in an 18—month beef system over a period of three years, using clover—rich swards or grass swards given 260 kg N/ha/annum. Despite a mean clover content of between 36% and 39% (in the DM) compared to less than 6% in the fertilized swards, there was no difference in daily liveweight gain (DLWG) between the two systems during the grazing season. However, during the fattening phase, the cattle fed on silage made from the grass/white clover swards had a 16% higher DLWG than those fed on grass N silage.

Clarke reviewed a series of beef cattle experiments and concluded that while stocking rates on grass/white clover swards were about 80% of those achieved when between 180 and 400 kg N/ha were applied, differences in individual animal grazing DLWG were relatively small, resulting in lower lwg/ha on clover-rich swards. More recently Young also reported that beef cattle grazing clover-rich swards, 36%—40% clover in the herbage dry matter, had a similar DLWG to cattle grazed on grass swards given 200 kg N/ha (Table 9).

Table 9. Daily liveweight gain (g/head) and liveweight gain/ha (kg) of beef cattle on grass/white clover and grass/N swards

	Grass/white clover		Grass/N	
	DLWG	lwg/ha	DLWG	lwg/ha
Steen and Laidlaw	1100	1113	1050	1459
Younie *et al.*	1095	831	1075	1056
Young	900	794	920	950

The effect of sward height of grass/white clover swards on grazing beef cattle performance was investigated by Moore *et al.* They found that steers grazed on clover-rich swards maintained at a sward height of 5 cm in early season had a significantly lower DLWG than steers grazing swards maintained at 7 cm or 9 cm. Daily liveweight gains were 0.60, 1.04 and 1.06 kg respectively. Similarly in the later part of the season, early August to late September, significantly higher liveweight gains were obtained from swards maintained at 9 cm rather than 7 cm, 1.11 kg/d compared to 0.85 kg/d. It was interesting to note that the more relaxed grazing treatments produced a mean output of 1265 kg liveweight gain per hectare over a three year period.

By contrast more consistent animal responses have been reported with sheep grazing grass-white clover swards. One of the difficulties reported with clover/ sheep systems was maintaining adequate clover contents under continuous sheep grazing. Orr *et al.* observed a progressive decline in clover content in clover-rich swards continuously grazed by sheep, over a three year period, when maintained at 3, 6 or 9 cm. In the first year the lamb liveweight gains from the grass/clover swards were 223, 268 and 295 g/d respectively compared to 260 g/d for lambs maintained on a grass sward given 420 kg N/ha and maintained at 6 cm.

Although the clover content of the 6 cm grass/clover sward declined to approximately 5 % in year 3, it maintained the same animal output per hectare relative to the sward given 420 kg N, 82%, as it had in year 1. The swards used in this experiment were based upon perennial ryegrass cv, Aberystwyth S 23 and grasslands Huia. Vipond and Swift evaluated a sheep system based upon perennial ryegrass cv. Condesa and Aberystwyth S S S 184 clover.

They observed that over a three year period the mean lamb liveweight gain was 1037 kg/ha compared to 975 kg/ha from a Condesa + 160 kg N/ha sward. They also observed that the mixture of tetraploid late-heading perennial ryegrass and small leaved clover enabled a clover content of 29 % to be maintained from mid to late season in year 3 despite being continually grazed throughout the growing season.

Wright *et al.* investigated the potential for combining cattle and sheep grazing to improve animal output from grass/white clover swards. They found that grazing by cattle in early season resulted in significantly higher clover contents than when grazed with sheep, 12.1% compared to 6.5% of the live herbage mass in July. When weaned lambs were then grazed on the swards the mean liveweight gains were higher on those previously grazed by cattle, than by sheep, 124 g/d compared to 101 g/d.

20

Uses of White Clover

The agriculture significance of white clover is well known and documented—rhizobial N-fixation capacity, soil improvement properties, excellent feeding value for animal production. These advantages have been neglected in Europe in the recent era of heavy fertiliser N application to grassland. However, the new European Community (EC) agricultural policies, e.g. EC Council Regulation No. 2078/92, aimed at environmentally friendly production methods and sustainable low input systems, will undoubtedly increase the exploitation of white clover (and other forage legumes); there is also similar burgeoning interest in non-EC countries.

Arable Cropping Systems

An important alternative use for white clover is as a support to arable cropping systems, mainly through its contribution of N. Various methods have been documented. While ploughing-in is a traditional way, more novel methods have been investigated, e.g. the direct drilling of maize into a perennial ryegrass/white clover mixture previously harvested for silage, with the regrowth between cultivated strips being cut and left as a mulch.

This process was repeated for three years and while maize yields were generally lower than those from traditional maize cropping, nitrate leaching losses were lower and there was less soil erosion or compaction; also, herbicide use was reduced.

Destruction of a white clover cover crop, e.g. by flaming, increased maize yields but at the expense of continuity of cover crop. Rototilling a clover mulch resulted in sweetcorn yields comparable to those from conventional cultivations but with less disease, while yields were lower with either unsuppressed or periodically mown clover.

Clover Understorey

Grain yields from the band-sprayed treatments were 60-78 per cent of those from drilling into bare land, but were substantially lower when drilled into the untreated clover understorey. When spring barley, spring oats and winter barley were grown in a white clover understorey, the yields of whole crop cereal for silage or of cereal grain were similar to those from traditional methods. In one experiment clover survived to be used again but in a second it was killed by a high slug population. Where clover survived, no weed, insect or disease problems were encountered.

Clements and Asteraki suggested that the low levels of pest attack could be the result of a build-up in clover of predatory insects and spiders which attack aphids and other pests; also, the emerging cereal crop may be camouflaged by clover against aphid attack. On account of the trampling effect, heavy sheep grazing after cereal sowing and again after cereal harvest was proposed as a means of slug control.

Set-aside Land

So-called set-aside land, i.e. land taken out of arable cropping because of overproduction, is assuming greater importance in EC agricultural policies e.g. EC Council Regulation 1795/92. A dense soil cover is advantageous in preventing soil erosion, leaching of N and weed invasion. In contrast to natural revegetation, a grass/white clover sward fulfils these objectives and soil N status is improved for subsequent cropping. White clover is particularly suited for long term set-aside.

Green Manure

Green manures in general have an obvious role in rotations -in organic systems in particular—and are of benefit to soil physical and chemical conditions. Advantages include augmentation of organic matter, accumulation of nitrogen, reduced soil erosion and nutrient leaching and a degree of weed control. White clover is generally considered to be shallow rooted in relation to red clover or to several grass species but nevertheless, some degree of recycling of mineral nutrients from the soil profile can be postulated.

In organic stockless system of arable cropping, leguminous mixtures including white clover as a component are typically sown as a green manure; the crop may be cut periodically during the growing season and the material left as a mulch. Specific uses of white clover as a green manure have been discussed by Barney and Ten Holte and Van Keulen; for example, in trials in The Netherlands, the latter obtained increased sugar yield and N uptake in sugar beet and increased tuber yield and N uptake in potatoes following the ploughing-in of clover sown under a winter wheat crop.

Horticulture and Viniculture

White clover, pure-sown or in association with grass, fulfils several functions as a soil cover crop or living mulch in various types of orchards, market gardens and vineyards although irrigation may be required during drought periods to aid its persistence. Engel listed several advantages of sowing legumes, including white clover, as: rapid soil cover, weed growth inhibition, improved soil structure, increased soil N and organic matter content, increased availability of nutrients; reduced soil surface temperatures in hot summers could also be added.

However, potential disadvantages were also noted: high seed costs, a haven for certain pests, e.g. rodents, increased risk of radiation frosts, excessive soil N enrichment leading to adverse effects on the chemical composition of the fruit. Tillage to establish cover crops may damage the roots of trees or vines but direct drilling (= slot seeding) offers an alternative method of establishment and has been successfully used to establish white clover in vineyards.

Extensification and Diversification

The wave of European extensification policies is resulting in positive measures to conserve plant and animal wildlife, and enhance landscape beauty, by more emphasis on floral diversity in swards. This diversity flourishes in low fertility situations but the cost of wildflower seed mixtures and the low output from such situations militate against the widespread creation of wildflower meadows. In Scotland, a number of commercial wildflower mixtures formulated by *The Nature Conservancy Council* were evaluated by Frame and Tiley.

Mixtures containing white clover (and other forage legumes) were more productive by up to 2.5 t/ha DM annually and richer in mineral concentrations than mixtures without forage legume components, yet a satisfactory floral diversity was maintained. They concluded that clover-containing mixtures offered a "middle way" between low fertility and herbage production from classical legume-sparse wildflower meadows and high production from intensively N-fertilised grass swards. The efficiency of this approach in extensive stock rearing enterprises was demonstrated by Fisher and Roberts for dairy youngstock.

In New Zealand a parallel approch by Ruz-Jerez *et al.*, compared a forage herb-rich mixture (of particular but not exclusive interest to organic farmers) with a conventional grass/white clover sward and an intensively N-fertilised grass sward (400 kg/ha/year) under sheep grazing. White clover made a substantial contribution to herbage production from both mixed-species swards and relative to production at 100 from the N-fertilised grass sward, the herbal ley and grass/clover sward gave values of 90 and 69 respectively.

Organic Animal Husbandry

Clearly white clover as a component of productive mixed-species swards plays an important part in organic animal husbandry and its products such as organic milk or meat, and several success stories have been documented. In his work on organic beef production, Younie noted the difficulty of maintaining the soil potassium (K) status under grass/white clover swards due to the removal of heavy silage crops, even altough animal manures were returned to the silage areas. Clover depletion in the swards could eventually result since clover's sensitivity to K deficiency is well documented.

Deer Farming

White clover is a valuable constituent of seed mixtures for deer farming. In a comparison of tall fescue versus perennial ryegrass swards in New Zealand, the tall fescue-based pastures had significantly higher carrying capacities because of a better clover balance.

Horses and Ponies

The number of horses and ponies kept for recreational riding has increased in recent years particularly in semi-rural areas. White clover is incorporated in special seed mixtures both to improve the nutritive value of the herbage and to enhance soil fertility.

Pig Feeding

There is some interest in the use of white clover as a feed for pigs, mainly because of its richness in protein. For example, in Switzerland Jost successfully fed grass/clover silage to pregnant sows.

Wildfowl

The wintering of migratory wild geese has led to increasing conflict with agriculture in various countries since sown grassland in certain localities is subjected to high grazing intensity during winter. The two main wild geese species in Scotland are the barnacle (Branta leucopsis) and the greylag (Anser anser). Experiments showed that their winter grazing substantially reduced the herbage available for subsequent spring grazing by cattle or sheep or for silage cropping. White clover is a "preferred" species by geese for grazing and use of this fact is made in refuges where grassland management is geared to maintaining clover-rich swards in order to attract geese from farmland. Many species of wild birds utilise clover as part of their diet, especially in winter but wildfowl are possibly the most significant with respect to sward damage; nevertheless, in autumn-established grass/clover mixtures bird damage to clover, e.g. by wild pigeons, can be severe.

Food and Wine

The importance of flowering white clover for honey bees and hence honey production is well known and documented. The honey bees are also responsible for pollination in clover seed crops. A clover-rich sward is encouraged and the clover allowed to flower in the avenues of a recently-established arboretum at The Scottish Agricultural College, Ayr, partly for the benefit of honey bees, but also from a landscape beauty point of view.

White clover is regarded as a plant edible by humans: "An infusion of the dried flowers makes a fine tea substitute. The young leaves, gathered before flowering, can be added to salads, sauces and soups? on their own they can be used as a vegetable and prepared like spinach." Moderation in use is recommended possibly because of the cyanogenic potential of some clovers varieties; Wheeler and Vickery found that north American varieties had notably lower cyanogenic potential than most European varieties. Again, it is perhaps not appreciated that a type of dry "wine" can be made by fermenting flowerheads, sugar and wine yeast.

Amenity and Reclamation

The stoloniferous growth habit of white clover makes it suitable for inclusion in seed mixtures for the stabilisation of slopes using various establishment techniques including conventional sowing methods, slot seeding or hydraulic seeding. Its N input is also valued for stimulating the growth of associated grass species and this is a major reason for including it in mixtures when reclaiming degraded land after opencast mining or mine spoils and tailings.

White clover is of value in various kinds of amenity or recrational grassland, partly for its agronomic properties, but also for its attractiveness when flowering. It could also be used more widely in gardens than at present, although its stoloniferous habit could be regarded both as an advantage, e.g. ground cover, and a disadvantage, e.g. its unwanted spread into neighbouring areas. In some countries, fourleaved clover is regarded as a good luck symbol and living plants may be sold or the leaves coated with a precious metal such as silver or gold and sold as jewellery.

Bibliography

Donnelly, E.D., and J.T. Cope, "Crimson clover in Alabama", *Bulletin* 335, Alabama Agricultural Experiment Station, Auburn, 1961.

Johnson, W.C., and J.I. Wear, "Effect of boron on white clover (Trifolium repens,L.) seed production", *Agronomy Journal,* 1967.

McCree, K.J. and J.H. Troughton, "Non-existence of an optimum leaf area index for the production rate of white clover grown under constant conditions", *Plant Physiol.*, 1966.

_____________ , "The prediction of growth rate from incident light or carbon dioxide uptake : a laboratory experiment with white clover", In : F.E. Eckardt (ed.), Functioning of Terrestrial Ecosystems at the Primary Production Level. *Proc. Copenhagen Symposium.* UNESCO, Paris, 1968.

_____________ , and S. Kresovich, "Growth and maintenance requirements of white clover as a function of daylength", *Crop Sci.,* 1978.

_____________ ., "Equations for rate of dark respiration of white clover and grain sorghum, as functions of dry weight, photosynthesis rate and temperature", *In: Crop Science,* 1974.

Mortvedt, J.J., and O.E. Anderson, "Forage legumes: diagnosis and correlation of molybdenum and manganese problems". *Southern Cooperative Series Bulletin* 278, Georgia Agricultural Experiment Station, Athens, 1982.

Rodgers, H.T., "Iron deficiency of crimson clover on a calcareous soil and method of diagnosis", *Agronomy Journal,* 1947.

_____________ ., "Boron response and tolerance of several legumes to borax", *Agronomy Journal,* 1947.

_______________ ., "Water-soluble boron in coarse-textured soils in relation to need of boron fertilization for legumes", *Agronomy Journal,* 1947.

Skipp, R.A.; Christensen, M.J., "Invasion of white clover roots by fungi and other soil micro-organisms IV", *Survey of root-invading fungi and nematodes in some New Zealand pastures, New Zealand Journal of Agricultural Research,* 1983.

Wear, J.I., "Boron requirement for crimson clover seed production, its accumulation in soils, and residual effects on sensitive crops", *Agronomy Journal,* 1956.

Wilson, D.R., C.J. Fernandez and K.J. McCree, "Carbon dioxide exchange of subterranean clover in variable light environments", *Crop Sci.,* 1978.

Index